The Nature of Florida's Ocean Life

Including

Coral Reefs, Gulf Stream, Sargasso Sea and Sunken Ships

written and illustrated

by

Cathie Katz

Atlantic Press
PO Box 510366
Melbourne Beach, Florida 32951

Other books by Cathie Katz

The Nature of Florida's Beaches
Including Sea Beans, Laughing Gulls and Mermaids' Purses

The Nature of Florida's Waterways
Including Dragonflies, Cattails and Mangrove Snapper

The Nature of Florida's Neighborhoods
Including Bats, Scrub Jays, Lizards and Wildflowers

The Nature of Florida's
Ocean Life

Including
Coral Reefs, Gulf Stream, Sargasso Sea and Sunken Ships

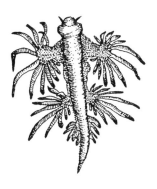

Written and Illustrated by Cathie Katz

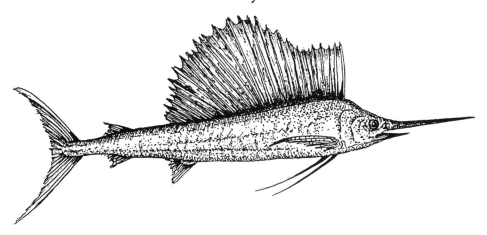

Atlantic Press
PO Box 510366
Melbourne Beach, Florida 32951

The Nature of Florida's Ocean Life
Including Coral Reefs, Gulf Stream, Sargasso Sea and Sunken Ships

Copyright by Cathie Katz © 1998
Photographs by Jim Angy © 1998
Poems by Patricia Ryan Frazier © 1998
Back cover art by David Williams © 1998

Published by
Atlantic Press
PO Box 510366
Melbourne Beach, Florida 32951

Printed in the United States of America.

All rights reserved. No part of this book may be reproduced or transmitted in any form or by any means, electronic or mechanical, including photocopying, recording or by any information storage and retrieval system, without permission in writing from the author.

Cataloging-in-Publication Data
　　Katz, Cathie
　　The Nature of Florida's Ocean Life Including Coral Reefs, Gulf Stream, Sargasso Sea and Sunken Ships

　　Includes bibliography and index.
　　p.　　cm.

　　1. Seashore Biology - Atlantic Coast. 2. Natural history - Florida. 3. Oceanography - Gulf Stream. 4. Oceanography - Sargasso Sea. 5. Marine invertebrates - Identification. 6. Marine Biology - Reefs. 7. Seed - Dispersal. I. Title.

ISBN Number: 1-888025-11-5　　　　Library of Congress Catalog Card Number: 97-077819

The Nature of Florida's Ocean Life
Contents

Introduction 2

I. OCEAN SHORES 8
Lighthouses 9
Cuban rafts 10
Cuban bottles 11
Sand 12
Life and death in the sand 13
Model homes 14
The best of all worms 15
Bioluminescence 16
Onions, walnuts and sloppy guts ... 17
Nurdles, duckies, and LEGO® toys .. 18
Lost soles 19
Inlets 20
Rock stars 21
Night creatures 22
Algae 23
The most abundant bird in the world ... 24
Ocean birds 25
A little whale stuff 26
Whale gut treasure: ambergris 27
The great eel mystery solved: migration .. 28
Eel life 29

II. REEFS 30
Sunken ships 31
Sunken Treasures 32
Human-made reefs 33
Coral reefs 34
Corals: not just another pretty face ... 35
Tongues, cameos and blennies 36
Sponges 37
Octopus and squid facts 38
Octopodes 39
Strange relatives 40
Reef bottom 41
Lobsters 42
Lobster migration 43
Shrimp 44
Reef at night 45
Reef worms 46
Fish 47

III. GULF STREAM 48
World travelers 49
The Gulf Stream: a little history . 50
Sea soup 51
Free spirits 52
Jellies 53
The drifters 54
Gulf Stream passengers 55
Fastest fish in the ocean 56
Disguises in the Gulf Stream 57

IV. SARGASSO SEA 58
Sargasso community 59
Sargasso: the ultimate cruise 60
Rain forests in the sea? 61
Sea bean soup 62
Fish in the sargassum 63
More fish in the sargassum 64
Designer creatures in the sargassum ... 65
Loggerhead sea turtles 66
For whom the light glows 67
Deep sea changes 68
Eels in the Sargasso Sea 69
El Niño 70

Bibliography 72

Index 73

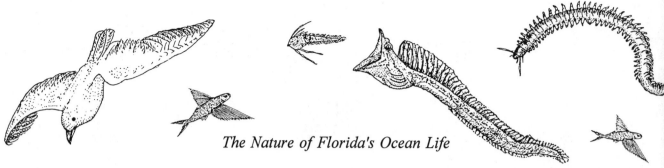

The Nature of Florida's Ocean Life

This book looks at Florida's ocean as a habitat, a community of elaborate mini-creatures swimming next to bulky saltwater giants. Drifting alongside these living creatures are inanimate objects -- bottles, nurdles, rafts, toys and tar blobs -- lost or tossed from cargo ships. They float in the Sargasso Sea or flow with the Gulf Stream -- sometimes drifting to Florida's coast. These objects have become as much a part of our ocean as fish and seaweed.

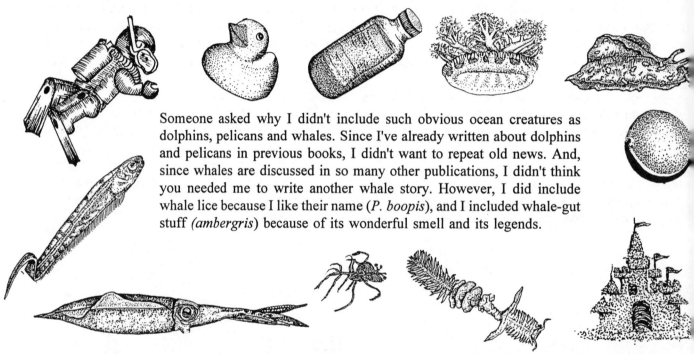

Someone asked why I didn't include such obvious ocean creatures as dolphins, pelicans and whales. Since I've already written about dolphins and pelicans in previous books, I didn't want to repeat old news. And, since whales are discussed in so many other publications, I didn't think you needed me to write another whale story. However, I did include whale lice because I like their name (*P. boopis*), and I included whale-gut stuff *(ambergris)* because of its wonderful smell and its legends.

When certain species are particularly exciting to me, like parchment-tube worms or sea beans, I've included them, even at the risk of repetition. But I try to describe their life at sea rather than at the end of their journey. For example, when I find a sea bean on the beach, it's just another inanimate, brown blob from who-knows-where. But when I realize this seed may have fallen from a South American rain forest into the Amazon River, then traveled to the Atlantic, floating in the Sargasso Sea in a thick, wet jungle of weeds, squid, shrimp, eels, slug eggs and sea turtles ... wow ... that's exciting. That's the life of Florida's ocean.

Shells, beans, weeds, corals, shoes and even humans are connected in this exotic community. To me, every rock, hole and odd-shaped clump has a story: Is it alive? Is it petrified? Is it toxic? Is it a worm? I'd like you to see their homes, neighbors, struggles at sea ... and discover how they survived (or didn't) and came to our shore.

I want you to taste the saltiness of this sea soup, feel the tentacles of spiny lobsters, smell the sharp odor of a sloppy guts anemone, see coral polyps coming out at night, and hear the loud pop of a pistol shrimp.

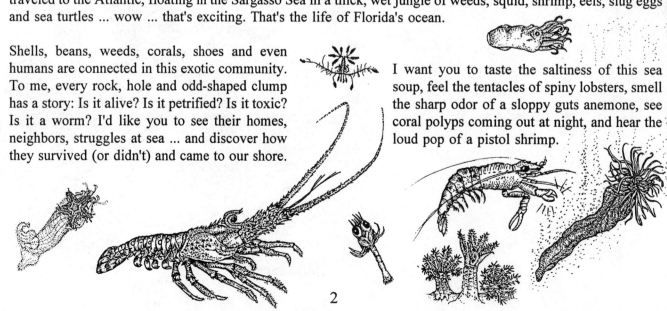

Introduction

My introduction to ocean life was from my mother's womb -- I went with my father and brother, and *in* my mother to Atlantic City, New Jersey in 1948. This must have been an impressionable time of development for me -- why else would someone born in Philadelphia and raised in the suburbs of South Jersey feel such a strong bond with the shore? When I was taken to Ocean City as an air-breather three years later, I *knew* I was home. The rest of my childhood didn't matter except when we vacationed at Ocean City, New Jersey or one of the beaches in Florida.

Since then, I've lived near oceans in the States, Europe and the Middle East, except for two years in the sixties when I worked in Wiesbaden, Germany. That period in the middle of a land mass was the most unproductive and uncomfortable time for me. This mid-continent life ended with a near-fatal crash on the Autobahn. After four months in the Kaiserslautern Hospital, I returned to the States -- to Lauderdale-by-the-Sea in south Florida. I healed by sitting in the surf, letting ocean nutrients swirl around my stitches and the wave pressure massage my broken bones. With the ocean's energy and healing powers, I mended enough to flex my muscles and move on. Since then, I haven't strayed very far from the ocean. It's where I belong. Biology or fantasy? It doesn't matter. When people say they "feel like a fish out of water," I know what they mean.

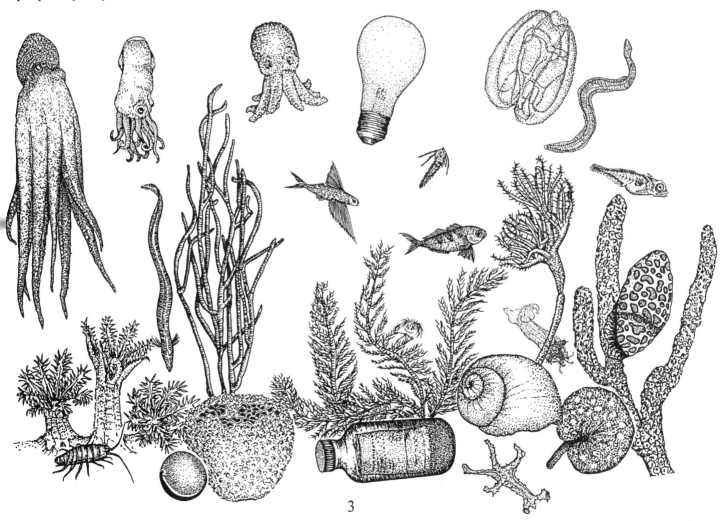

Introduction

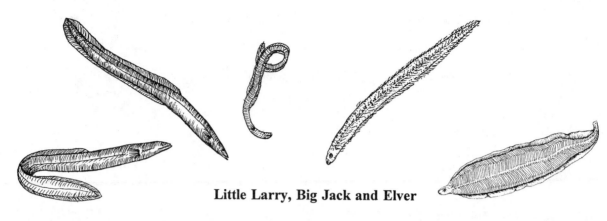

Little Larry, Big Jack and Elver

Once again, Little Larry has forced her tiny stick body onto every page in this book -- splashing, laughing, pushing and plunging where she doesn't belong. With her is Big Jack, the shy boy who washed ashore on a sea bean last year. He is the namesake of Jack Hoskins, our lapidist in Melbourne who helps beachcombers polish rocks and sea beans. Both Big Jacks are gentle, wise and serious, and both tolerate Little Larry's odd behavior. The only difference between these Jacks is that one is less than half an inch tall and the other is way over six feet.

In this book, Little Larry and Big Jack go to the inlet to look for sea beans. Instead, they discover Elver, an immature eel who, *as they can clearly see*, is European. But how did a European eel cross the Atlantic to end up in Florida's waters?

American and European eels leave their inland waterways to migrate to the Sargasso Sea to reproduce. After their offspring are born, the parents die without ever returning to their native continent. Their babies, as tiny as sesame seeds and as clear as glass, grow into *elvers*, a phase between newborn and adult. American elvers swim to America to begin life as freshwater fish, while European elvers head in the opposite direction. These elvers have no parents to lead them, but a strong instinct guides them to their ancestors' homeland. However, one of these immature eels, Elver, apparently lacks this sense of direction, and finds himself at a Florida inlet.

Little Larry and Big Jack decide it's their mission to guide Elver toward his homeland in Europe. Each page shows their journey as the three of them meet morays, anemones, ghost shrimp, worms, squid, Cuban rafts and barracudas. As they cross dangerous reefs, powerful Gulf Stream currents, and jungles of the Sargasso Sea, they discover a world of surprising connections.

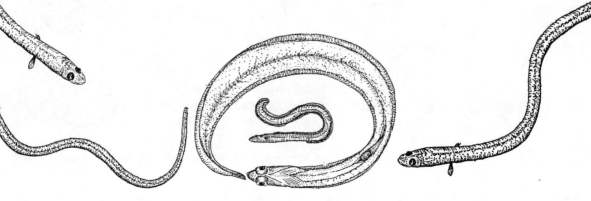

"The planet earth is misnamed, if you consider it from the proportion of land to water. By a fair naming process it would be called planet water."
- John and Mildred Teal in *The Sargasso Sea*

♪♫♪ This Magic Moment ♪♫♪

This book is dedicated to The Drifters, a magical group of beachcombers, scientists and naturalists who love ocean life in all its forms -- from messages in bottles and yellow plastic duckies to sea beans and sea weeds. We communicate through a newsletter called *The Drifting Seed* and meet once a year in Melbourne Beach at the International Sea-Bean Symposium to exchange discoveries, ideas and treasures.

We are from Kenya, Ghana, Zambia, South Africa, Tasmania, Australia, New Zealand, Japan, Korea, Malaysia, Guam, Austria, France, Switzerland, Germany, Belgium, Holland, the United Kingdom, Canada, the United States, Mexico, Venezuela, Peru, and most recently, the Russian Federation.

Ever since we united through *The Drifting Seed*, The Drifters have inspired me to a happiness and joy that can only be given by people who love what they do and who cherish what they see in the natural world. They showed me how all of us and how all things are connected.

Dr. Bob Gunn and John Dennis, co-authors of *The World Guide to Tropical Drift Seeds and Fruits*, are my mentors and the original magic men. This book is for them and The Drifters who came together because of their work.

The Drifters

Jim Angy from Indialantic, Florida
Kenn Arning from Seattle, Washington
Marge Bell from Melbourne, Florida
Scott Boykin from Gulf Port, Florida
Sue Bradley from Melbourne Beach, Florida
Dr. Gerhard Cadée from Texel in The Netherlands
Dr. David Cox from Vero Beach, Florida
Rita Crane from Dunedin, Florida
John V. Dennis, Sr. from Princess Anne, Maryland
Dr. Curtis Ebbesmeyer from Seattle, Washington
Dr. Charles R. Gunn from Brevard, North Carolina
Tim Kozusko from Cocoa Beach, Florida
Dr. Stephen Leatherman ("Dr. Beach") from Miami, Florida
Hiroki Nakanishi from Nagasaki, Japan
Dr. Charles Nelson from Wisbech, Norfolk, United Kingdom
Pamela Paradine from Somerset, United Kingdom
Ann and Ian Robertson of Malindi, Kenya
Janice Scott from Cocoa Beach, Florida
Frank Seymour from Jensen Beach, Florida
Ruth Smith from Arlington, Virginia
Rhonda and Dean Theobald from Merritt Island, Florida
Xander van der Burgt from Arnhem, The Netherlands
David Williams from Midlothian, Virginia
Cathy Yow from Houston, Texas
Peter B. Zies from Lake Mary, Florida

...and more than 250 others who have magically come together.

"The meeting of two personalities is like contact of two chemical substances: if there is any reaction, both are transformed."
- Carl Jung

K. P. - OCEANSIDE

Mermaids never wash their dishes,
they simply toss them away!
The ocean scours them with sand,
a scrub with seafoam,
a rinse with spray.

Beachcombers find the discards
on every shore and nook;
While mermaids munch
their shellfish meals
without a second look!

- Patricia Ryan Frazier

"The value of science remains unsung by singers, so you are reduced to hearing — not a song or a poem, but an evening lecture about it."
- Richard Feynman, U.S. physicist

I hope this book serves as song for all the ocean scientists.
- Cathie Katz

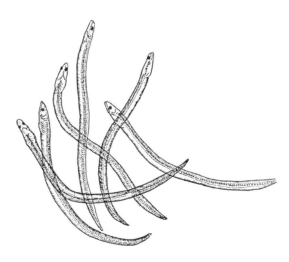

The Nature of Florida's Ocean Life

OCEAN SHORES

"If there was a big bang the universe must have consisted of an infinite amount of energy concentrated in a single point. God knows where that came from."

- Gerhard Staguhn, U. S. physician

Lighthouses

High Lights
Starting in the 1820s, more than 30 lighthouses were built in Florida in response to increased ocean traffic, piracy and too many ship wrecks on Florida's treacherous reefs.

Near my office at Cape Canaveral Air Station is a black and white banded lighthouse built in 1847. When I stand at the top of the 139-foot tower, I can see the ocean's coastline and the entire Cape with all its gantries, assembly buildings and retired rockets jutting from the swamps and scrub fields.

Protection or Sabotage?
Lighthouses were supposed to protect ships, but during World War II, enemy submarines prowled in deep Florida water, stalking tankers that were illuminated by the lighthouse's beam.

"The area around Cape Canaveral was a very heaven for submarines, located as it was on one of the most deserted stretches of the seaboard ... March 1942 saw a decline of undersea attacks, but April and May sinkings multiplied until this period became the most disastrous of the entire war along the Florida East Coast."
- A.J. and K.A. Hanna in *Florida's Golden Sands*

After a German submarine torpedoed the U.S. tanker *Pan Massachusetts* off the Cape's coast, the lighthouse was ordered to stay dark until the war ended.

LEGO Lighthouse?
Severe beach erosion forced the designers of the Cape Canaveral lighthouse to create a relocatable structure which hadn't been done before. They made removable cast iron plates for the body, a detachable spiral staircase and a brick-lined interior that fit together in pieces.

The original structure can now be dismantled and re-built for a fraction of the cost of building a new structure. In 1893, severe erosion threatened the lighthouse's stability, so the innovative structure was put to the test -- and passed. It was successfully dismantled and moved, then re-assembled at the spot where it still stands today.

"I joined the ranks of lighthouse lovers in 1953 when I first visited the Cape Canaveral Lighthouse. Our detachment of rocket scientists, engineers and technicians arrived at the Cape to prepare and launch the first of a long series of Redstone rockets ... I immediately felt its ties to the past and was excited to climb the circular staircase and look over the vast expanse of the Cape. Our director, Dr. Wernher von Braun was stationed in the balcony of the lighthouse to observe the rocket launches from Pad 4."
- Frank Childers in *History of the Cape Canaveral Lighthouse*

Cuban Rafts

Since 1959 when Fidel Castro took power in Cuba, more than 700,000 Cubans have come to settle in Florida. Many escaped Castro's island by going across "el Charco" (the puddle), the treacherous 90-mile stretch of water, in homemade boats.

While beachwalking in the summer of 1994, I found about a dozen battered rafts washed up with the high tides. Each was covered with seaweed and beach junk. The rafts had rudders made from school desk lids, hulls from steel drums, masts from sign posts, seats from worn tires ... evidence of thousands of people desperate to leave Cuba. Some of the rafts I found were marked with neon spray paint *"CG OK"* (*"Coast Guard found passengers: they are OK"*), signalling the Coast Guard had taken the occupants on board. But what about the rafts that weren't marked? Where were those owners?

Discovering a raft while beachwalking is different from finding a sea bean, shell or piece of driftwood. Humans had drifted on the ocean for miles in that crowded and flimsy sailing vessel.

Where were they? Were they alive? Finding the raft makers seemed impossible, but when I closed my eyes, I envisioned anxious bodies drifting toward Florida's shore. Did I want to make a connection so bad that I was able to conjure images of still-alive humans?

In 1994, I found a plastic-wrapped package half-buried in the sand. Tightly bound in plastic were the paper records of a young Cuban man named Ernesto. His driver's license, government documents, and addresses in Miami gave me clues to one of the raft's owners. I contacted each address in Miami, with no luck until one gentleman said he knew of Ernesto, but didn't know about his departure from Cuba. I sent him copies of the paperwork anyway, with a request to call me if anyone heard from Ernesto. I had no hope of hearing anything more.

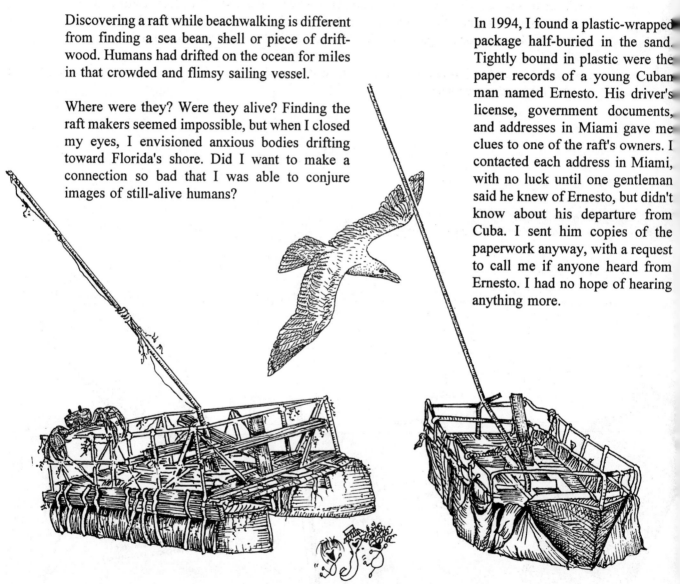

I just called to say hallelujah!

Eleven months later, on July 4, 1995, I received a call from Ernesto's uncle. In the background I heard, "Hallelujah! Hallelujah! It's a miracle!" It was Ernesto, my lost young Cuban, finally in Miami after being detained in Guantánamo Bay for almost a year. Ernesto had dropped his package of identity papers when he climbed on board the Coast Guard vessel that had carried him to Guantánamo Bay. Ernesto wanted me to know he never lost hope of getting to Florida.

"One does not discover new lands without consenting to lose sight of the shore for a very long time." - André Gide

Cuban Bottles

During the 1990s, I found a half dozen six-inch brown bottles in the wrack line of seaweed and debris; each bottle had a card inside like the one shown here. For the next few years, hundreds of other beachcombers from the Keys to northern Florida continued to find these brown bottles.

Florida Today Newspaper, May, 1990:

Nothing sinister in those Cuban message bottles

"In all, 14,930 bottles have been cast into the sea from Cuba in the past three years as part of an experiment to track ocean currents in case of an oil spill, a Cuban government spokesman says ... The message encouraged speculation about CIA plots, Cuban exile maneuvers and neighborhood pranks ... Cuban Special Interests Section in Washington said the Cuban institute began dropping the bottles in 1988 to track currents."

Like hundreds of other Cuban-bottle-finders, I wrote to the institute asking for information about the drift bottle study, but none of us ever received a response.

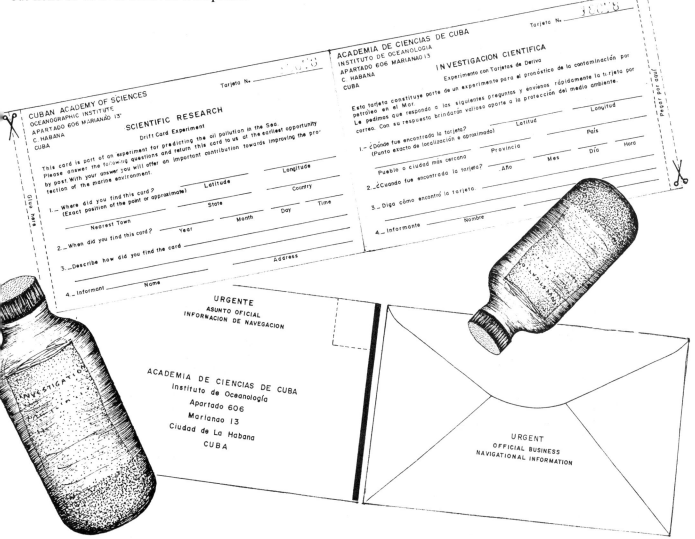

Cuba to Florida: 1898 to 1998

"With the end of the war much of the Cuban population of Florida returned home to share in the organization of the new independent government ... Since then the geographical barriers separating Florida from Cuba have become increasingly nominal. Many interchanges have developed on foundations laid in those dramatic years when Florida's coast played a part in Cuba's war for independence."
  - A. J. and K. A. Hanna (writing about the end of the Spanish-American War in 1898) in *Florida's Golden Sands*,

Sand

Is sandpaper really made of sand? Yes, sand grains are embedded onto glued stiff paper. The coarseness is determined by the size of grains that will fit through mesh screens. The reddish-brown color comes from garnet particles.

Why does the sand on Florida's beaches vary so much in texture and color? Most of our beaches are made of white sand that washed down from the Appalachian Mountains. Over millions of years, wind, water, temperature and Earth's movement can change a boulder to a grain of sand. Sandcastles are now built from these Appalachian boulders.

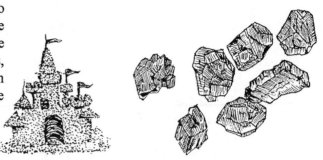

Colorful and textured sand on some Florida beaches is made from crushed shell material.

Glistening white sand along the Gulf of Mexico is made of smooth hard particles of limestone called *oolite* which means "egg stone" in Greek.

In the southeast, the chalky white sand is calcium carbonate from marine animals and plants.

Pure Sand?

"Panama City Beach sand is like a bar of Ivory Soap - it's ninety-nine and forty-four one-hundredths percent pure quartz. Quartz, the hardest component of beach sand, came from the Appalachian Mountains millions of years ago. Over the centuries, everything softer has been ground up and washed away, leaving feather-soft, blindingly white, quartz powder."
 - Dr. Stephen Leatherman ("Dr. Beach"), coastal geologist and Director of the International Hurricane Center at FIU in Miami.

Dr. Beach, author of *America's Best Beaches*, also says, "Florida is the best state for beaches, but the only *perfect* beach is a state of mind."

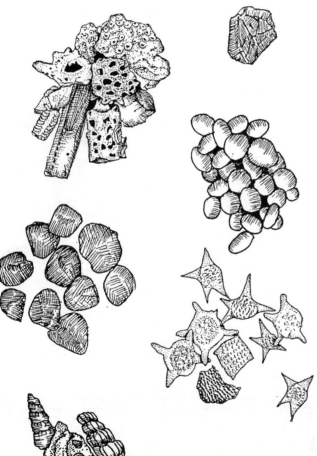

The drawings shown here are various sands magnified 30 to 100 times.

Green tinges in moist sand is caused by a zoo of microscopic animals called *meiofauna* that live in the spaces between grains.

"... sands are primarily derived from limy skeletons of sea creatures ranging from coralline algae to clams. Unlike the tawny material of similar texture that covers northern shores, this semitropical sand is, when seen under the microscope, instantly recognizable as chips and flecks of once-organized beings. Magnified to the same degree, a pinch of New Jersey sand is a lifeless-looking collection of irregularly worn pebbles."
- John L. Culliney in *The Forests of the Sea*

Life and Death in the Sand

A lot of sand is made from crushed material from the coral reefs. Corals, from their beginning are attacked, chipped and eroded by animals and plants. Sponges and mollusks destroy corals by secreting acids, and parrot fish and sheepshead scrape and crunch on coral tips, eating the algae and discarding the coral bits.

Coquina rock is formed when particles of sand are trapped in crevices within bits of coral and rock. A reaction occurs when an organism, such as algae, grows over the trapped particle. The chemical reaction, in combination with constant pressure of tides, seals the new mass together. The tides continue bringing in bits of shell and coral, constantly adding to this welding process until "rock" is formed.

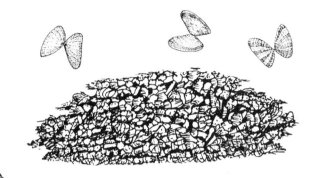

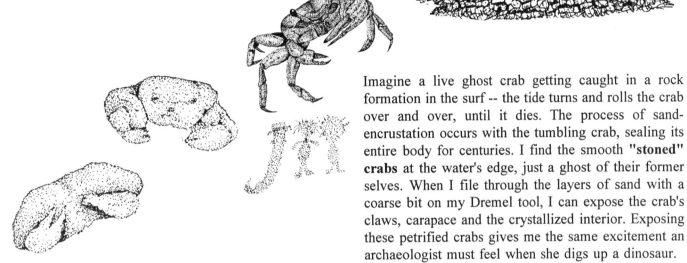

Imagine a live ghost crab getting caught in a rock formation in the surf -- the tide turns and rolls the crab over and over, until it dies. The process of sand-encrustation occurs with the tumbling crab, sealing its entire body for centuries. I find the smooth **"stoned" crabs** at the water's edge, just a ghost of their former selves. When I file through the layers of sand with a coarse bit on my Dremel tool, I can expose the crab's claws, carapace and the crystallized interior. Exposing these petrified crabs gives me the same excitement an archaeologist must feel when she digs up a dinosaur.

Ghost Poop

Some of the holes that I find in the muddy sand around tide pools are made by creatures I rarely see. **Ghost shrimp** (*Callianassa atlantica*) leave little black pellets that look like chocolate sprinkles outside their burrows. Next time you notice this pile of poop around a neat little hole, you'll know a ghost shrimp is living down below.

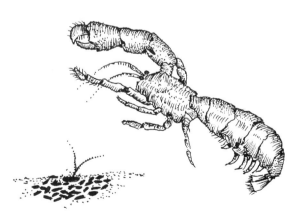

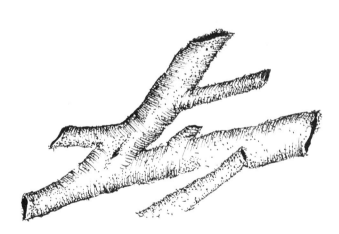

What happens when lightning strikes the beach? The sand melts and forms hollow tubes called **fulgurites**. Pieces of these fulgurites are hard to find because they look like sand, but if you ever see glass-hard tubular pieces, they might be evidence of lightning strikes.

"Mama exhorted her children at every opportunity to 'jump at de sun.' We might not land on the sun, but at least we would get off the ground."
- Zora Neale Hurston

Model Homes

Thousands of **honeycomb worms** (*Phragmatopoma lapidosa*) build colonies of tunnels from sand grains. They live inside, decorating the walls of their living quarters with bits of shell. Each worm is a long, thin thread that breaks easily if you try to pull it from its hole. During their life, they never leave their home so they're totally dependent on food washing over them as seawater flows by. Iridescent golden bristles on their heads emerge from their holes, adding a shining touch to the otherwise drab worm. Sometimes large chunks of these colonies (as shown to the right) break off and wash up with strong tides.

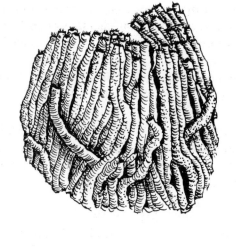

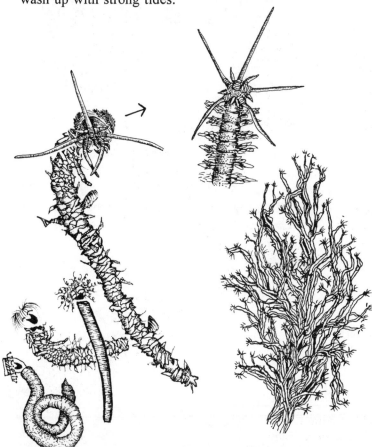

Another sign of life under the sand is the tube of a **plumed worm** (*Diopatra cuprea*). They are also called shaggy parchment tube worms because of the messy tube casings they make from grains of sand, ocean debris, shell pieces and mucus. The result is a long casing (in which they live) with all kinds of good junk stuck to it as shown to the far left. I see the ends of the tubes sticking out from the sand, looking like little plant stumps.

The tubes of **sea frost** (*Filograna huxleyi*) (shown left) look like a mass of tangled pale plants, but the twinings are actually the homes of common worms. They normally grow in a disorderly tangle of branches, but sometimes I see a little clump encrusted on a thick piece of rope or sponge. When alive, they poke their tiny heads out, but quickly retreat when I bend down to touch them.

Elephant tusks (*Dentalium eboreum*) are the shells of marine snails that live in the sandy bottom, eating their way through the mud. They feed by picking up particles with their sticky thread-like tentacles. As the sandy food passes through their body, nutrients are filtered from the sand particles, and the remainder is eliminated.

Tusk shells found by archaeologists in inland burial plots provide evidence that humans used tusk shells for trading as far back as 20,000 years ago.

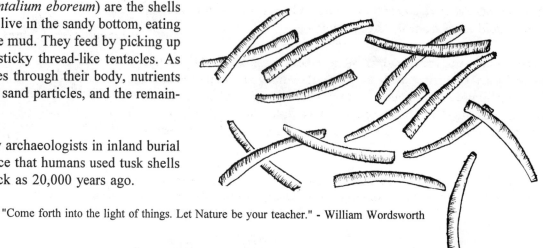

"Come forth into the light of things. Let Nature be your teacher." - William Wordsworth

The Best of all Worms

The **parchment tube worm** (*Chaetopterus variopedatus*) lives its entire life inside a papery tube made from slime secreted from one of its special glands. I often find the dried up tubes on the beach, but I had never actually seen the resident worm until the winter of 1998 when Rhonda Theobald and I were beachwalking by Cape Canaveral lighthouse. Rhonda picked up a particularly squashy tube and said, "I think there's a worm inside." I couldn't believe it, but sure enough, inside was a yellowish, soft worm body with legs, flippers and wings. She later preserved it and donated it to use in my beach talks.

I love this worm. When alive, it paddles, splashes, filters and lives in a homemade sack with a couple of pea-sized crabs. Its head looks like a toilet seat, it has suckers to anchor itself to its house walls, it has a roller-coaster system of getting food to its mouth, AND it glows with a bluish-green luminescence when you touch it. Wow. What more could anyone ask for in a worm?

MacPellets

This flamboyant animal has a "head" with mouth, tentacles, bristles, and two "wings." Its long body has five segments with three pairs of paddles in the middle. Both ends of the tube are narrow and stick out of the mud to allow water inside. The worm's paddles drive water through the tube, carrying oxygen and nutrients. The worm uses an intricate process to make a meal; it creates pellets from the ocean's particles; these pellets slide down a special groove to a mucous sac and then into its mouth.

I'm in awe of this worm because it can regenerate itself within two weeks from almost any part of its body. In a laboratory test, a whole new worm grew from just a middle portion of its paddles.

How do they reproduce if they never leave their tube? Each worm releases eggs and sperm into the water where fertilization takes place. When one worm spawns, it sends out a chemical signal which triggers all the others to spawn at the same time.

Save the Worm

Why am I so attracted to worms? Because they're night creatures, like me, and even though they're solitary they live near other worms. They are also deeply affected by the pull of the moon. I especially like them because they're squirmy, flexible and have lots of information that I could learn from if I'd pay attention.

When people ask rhetorically, "What kind of animal would you like to be?" I never answer with the more glamorous and expected animals like dolphins, eagles, wolves or sharks ... I always feel connected to worms. Happily, my DNA seems to have linked me with slimy, squirmy creatures rather than the more traditional symbols of nature.

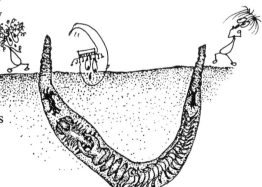

"Organisms fashion all these bio-minerals with great precision and under conditions that would make many processing engineers envious. Over and over they demonstrate that whatever engineers can do, nature can do better and on a much finer scale."
    - Elizabeth Pennisi in "Natureworks" in Science News

Bioluminescence

A still-unsolved mystery about some marine worms is their ability to *glow* when touched or disturbed. If the whole worm is handled, the entire body lights up.

Observing these bright creatures in their natural conditions is difficult because of the depth and the darkness. Some underwater researchers have witnessed deep sea creatures using light displays to identify themselves to willing females. Other creatures may use luminescence as an alarm to distract predators. Parchment tube worms have no eyes and they live their entire lives in a tube in the mud. And yet they glow. Why?

Dr. David Cox of The Environmental Learning Center in Vero Beach asks, "Are we suffering from a need to find a purpose in everything?"

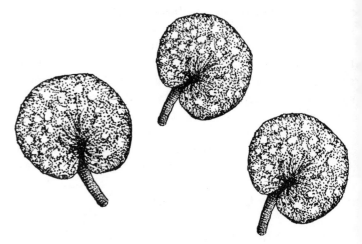

Shaped like a lily pad on a stalk, the beautiful purple **sea pansy** (*Renilla reniformis*) is another example of bioluminescence. When I found my first sea pansy on the beach I thought it was some kind of mushroom. But it's actually a spongy plant-like coral, classified as an octocoral. They have a high concentration of *luciferin*, an enzyme that makes them bioluminescent.

Sea pansies grow over some areas of the ocean floor as densely as a field of pennyworts. Sea slugs (*Arminia tigrina*) control the spread of these pansies by constantly grazing on them.

Other octocorals include **slimy sea feathers** (*Pseudopterogorgia americana*) which are mucous-covered strands of feathery branches; **sea whips** (*Leptogorgia virgulata*) which look like candy-covered wire branches of purple, red, orange, yellow or white; and **sea pens** (*Pennatula aculeata*) which remind me of frilly seaweeds. What an unusual group of creatures to be related.

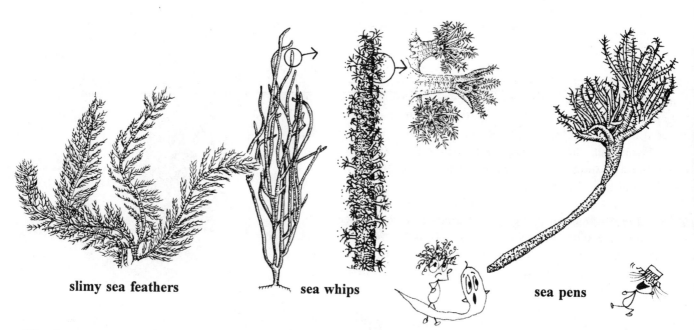

slimy sea feathers   **sea whips**   **sea pens**

"Phosphorescence! Now there's a word to lift your hat to! To find that phosphorescence, that light within; *that's* the genius behind poetry."
- Emily Dickinson

Onions, Walnuts and Sloppy Guts

Until I was about 12 years old, my family and I would spend summer vacations in Ocean City, New Jersey. As soon as we arrived, we'd scatter, like ants freed from a jar, to where we fit best: my father settled, just for the weekends of course, with his newspaper and cigar; my mother scurried to unpack, organize, and begin her busy nesting rituals; and my brother, daring and brave, bolted off to dive head first into the big waves. I melted into the soft wet shore, digging for worms, squeezing anemones, and playing with whatever slimy blobs I could find. Almost four decades later, we still show these same tendencies.

Nowadays in Florida, I love when **onion anemones** (*Paranthus rapiformis*) roll onto our shore. They look remarkably like our familiar edible onions. When I touch their watery, inflated bodies, they withdraw their tentacles so fast -- I don't get a chance to inspect them.

In 1997, Florida's beaches were invaded by hundreds of vegetables including onions: did they fall off a transport ship? Where did they come from? Will we ever know?

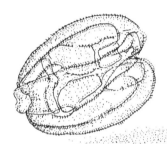

Another wonderful anemone is called **sloppy guts** (*Ceriatheopsis americanus*) (shown below). These anemones secrete gooey, greenish-brown mucous tubes in which they live. These tubes look like intestines; hence their name. Trying to dig them out of their sandy bottom habitat brings out their strong acid odor.

(More anemones are described in the next section, Reefs)

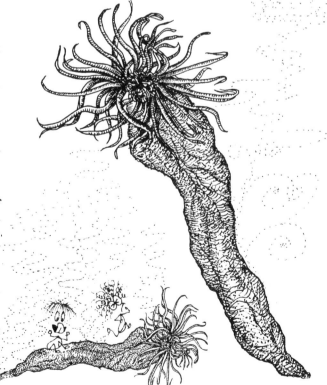

Did you ever notice round, yellowish "stains" on the beach near the tide line? They all seem to be about the same size. What creates these stains? **Comb jellies** (*Mnemiopsis macradyi*), shown above, float up and down in the sea water, drifting with the currents. Once in awhile they wash onshore. Because they're almost entirely made of water, they dissolve on warm sand, leaving behind only a tinted memory of their former beautiful shapes.

They swim near the water's surface, using rows of tiny hairs as paddles. These hairs are iridescent by day and luminescent by night.

What's the difference between iridescence and luminescence? Simply, iridescence is a *sheen*, as on a soap bubble, from colors playing on a smooth surface; luminescence is a *glow*, as in fireflies, that comes from a chemical action creating light.

"The simpler we make our lives, the more abundant they become." - Sarah Ban Breathnach

Nurdles, Duckies and LEGO® Toys

Dr. Curtis Ebbesmeyer, an ocean research scientist from Seattle (also known as "a filter-feeder for information about floating objects") has studied things that drift in ocean currents, from dead human bodies and Nike sneakers to nurdles and yellow plastic duckies. He says, "The literature of things that float on the ocean is so scattered that it doesn't make sense until you compress it all, then it begins to take on a glow, like radium."

When I asked Dr. Ebbesmeyer, "Why do tar, toys and sea beans arrive together on Florida's east coast?" he replied: "I think that the Gulf Stream transports toys, tar, seeds, bottles and myriad other items all intermingled in an immense river of flotsam. Beginning in October, the Stream and the winds push the stream of wrack westward touching Florida's coast, thereby dumping on the sand, dollops of the floating bouillabaisse."

What are **nurdles**? They are lentil-sized plastic pellets of different colors, transported in large bags to manufacturers who melt them down to make plastic products. Billions of them float in our ocean. So who lost their nurdles? We don't know. No one's talking.

Wanted: Yellow Plastic Duckies

In 1992, 29,000 plastic yellow duckies fell from a container ship leaving Hong Kong for Tacoma, Washington. When the ship hit rough seas, it listed with such force that the securing cables snapped from the strain, sending crates overboard. These duckies have since been tracked by Curt Ebbesmeyer and James Ingraham and found on beaches in Alaska, Oregon and Washington.

Some of the duckies are probably now frozen in arctic ice packs, drifting east with the transpolar current. Who knows? When the ice thaws, the duckies may head into the Atlantic. Florida's beaches may someday play host to these bobbing castaways. If you find a yellow duckie, write to me or Dr. Ebbesmeyer (address below). But watch out for impostors. The real duckies have "The First Years" imprinted on their side. Traveling with the duckies are blue turtles, red beavers, and green frogs. Ones found in other parts of the world have already become collector's items. Remember: beware of impostors. (An impostor duck is shown below left.)

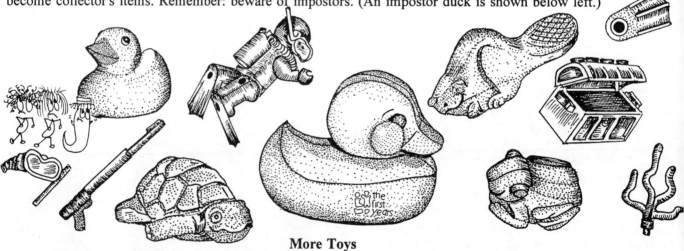

More Toys

Florida's coast will probably receive the contents from several other cargo spills: In February, 1997, the *Tokio Express* lost five million tiny LEGO® toys. Ironically, the lost toys include sea creatures, divers, life rafts and diving equipment such as tiny little flippers and masks. And in March of 1997, the *Cita* ran aground on Saint Mary's Island off the southwest coast of England and lost crates of toilet seat lids, hubcaps, car tires, body surfing boards, computer mice, Actionman kits, French wine, golf bags, sneakers, dress shoes, plywood and, among many other items, one million plastic shopping bags. The plastic bags, from an Irish supermarket, have this message printed in green: *Help Save The Environment*.

Anyone finding the above items, please contact **Dr. Curtis Ebbesmeyer, 6306 21st Avenue NE, Seattle, WA 98115**.
Dr. Ebbesmeyer produces a newsletter called *Beachcombers Alert!* which has all kinds of information about things that drift in the ocean. He is also author of *All Things That Float* and is interested in hearing about unusual items found on the shore or in the ocean.

Lost Soles

Whatever I find on the beach, excites me, whether it's a ping pong ball, sea turtle hatchling, lump of wax, shoe or glove. When I find *two* of the same item, I really get excited, and that's when I turn to my associate Drifter Dr. Curtis Ebbesmeyer for advice. Curt uses oceanographic studies, computer-generated programs, trajectory plots, phone calls to manufacturers, and lots of beachcombers' intuition to determine the origin of his finds.

According to Curt, "It's possible that during the summer of 1998, LEGO toys, toilet seats and flip flops may wash up on Florida's beaches. Remember the ocean axiom: *Never overlook anything on the shore*."

When 80,000 Nike sneakers spilled overboard, beachcombers on the Pacific coast found one or two at a time, happy with their never-worn (although wet) sneakers. But how well do sneakers wear in the ocean? Curt answers, "A Nike sneaker which, despite three years in the corrosive sea, remains wearable and the label readable enough to trace its serial number to their shipment."

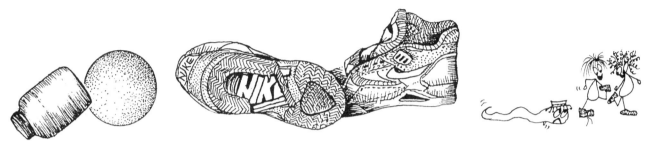

♪ ♫ ♪ ♫ ♪ ♫ ♪ Match Maker, Match Maker, Make me a Match ♪ ♫ ♪ ♫ ♪ ♫ ♪

But what do you do when your sneaker doesn't have a mate? Creative beachcombers in Oregon hold swap meets to search for mates. At the rate sneakers are washing overboard from cargo ships, Florida might be wise to pay attention to the soles of found sneakers: if the soles aren't worn, it's probably a new shoe and more will be on the way. Write down the serial number and be prepared to find your mate.

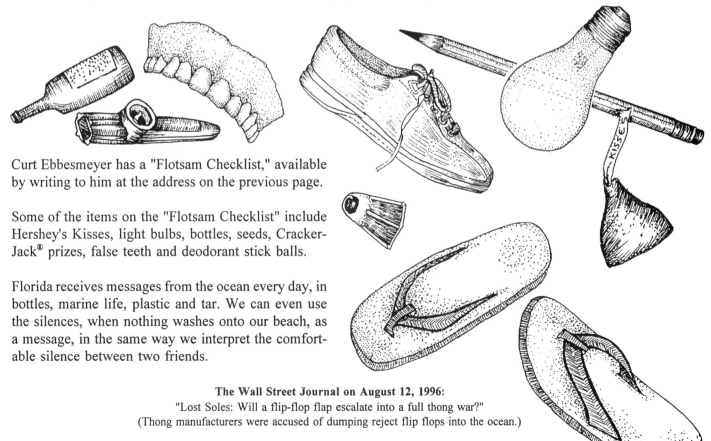

Curt Ebbesmeyer has a "Flotsam Checklist," available by writing to him at the address on the previous page.

Some of the items on the "Flotsam Checklist" include Hershey's Kisses, light bulbs, bottles, seeds, Cracker-Jack® prizes, false teeth and deodorant stick balls.

Florida receives messages from the ocean every day, in bottles, marine life, plastic and tar. We can even use the silences, when nothing washes onto our beach, as a message, in the same way we interpret the comfortable silence between two friends.

The Wall Street Journal on August 12, 1996:
"Lost Soles: Will a flip-flop flap escalate into a full thong war?"
(Thong manufacturers were accused of dumping reject flip flops into the ocean.)

Inlets

Inlets or outlets? These passageways allow treasures to pass between land and sea. An exchange of gifts -- coral, sea beans, jellies, spirula, bottles, sea whips, elvers, seaweeds -- some stay behind, others return. Outside the inlets, coral reefs display their own treasures while farther out, the fast water of the Gulf Stream and the whirling playground of the Sargasso Sea hint at exotic treasures I haven't even seen yet.

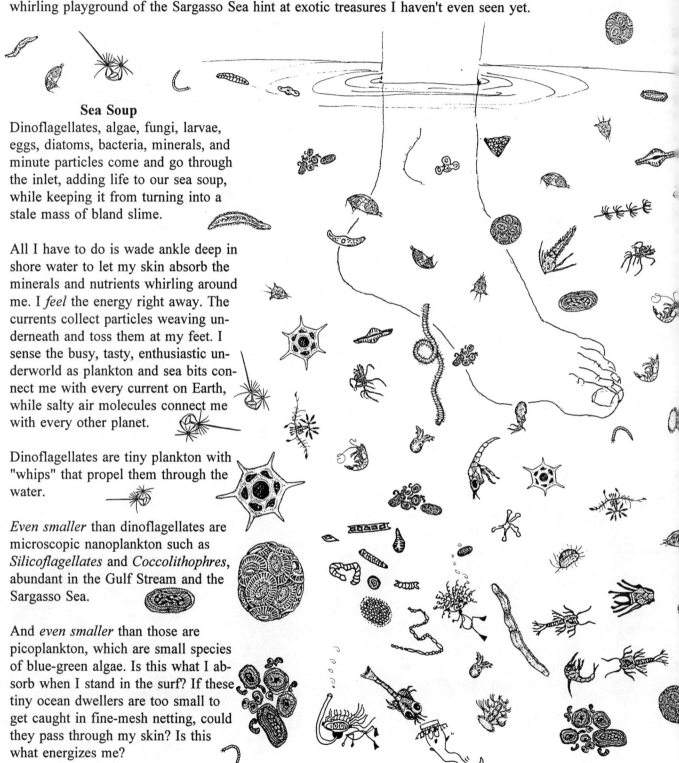

Sea Soup

Dinoflagellates, algae, fungi, larvae, eggs, diatoms, bacteria, minerals, and minute particles come and go through the inlet, adding life to our sea soup, while keeping it from turning into a stale mass of bland slime.

All I have to do is wade ankle deep in shore water to let my skin absorb the minerals and nutrients whirling around me. I *feel* the energy right away. The currents collect particles weaving underneath and toss them at my feet. I sense the busy, tasty, enthusiastic underworld as plankton and sea bits connect me with every current on Earth, while salty air molecules connect me with every other planet.

Dinoflagellates are tiny plankton with "whips" that propel them through the water.

Even smaller than dinoflagellates are microscopic nanoplankton such as *Silicoflagellates* and *Coccolithophres*, abundant in the Gulf Stream and the Sargasso Sea.

And *even smaller* than those are picoplankton, which are small species of blue-green algae. Is this what I absorb when I stand in the surf? If these tiny ocean dwellers are too small to get caught in fine-mesh netting, could they pass through my skin? Is this what energizes me?

"These tiny plankton, whose extent and importance were unimagined five years ago release one-third of the atmosphere's total supply of dimethyl sulfide. Coral reefs supply another third. Dimethyl sulfide, it turns out, is a condensation nuclei essential to cloud formation.
- William Thomas in "Sea Change" (1998)

Rock Stars

Some residents live in on rocks where seawater creates a drama of waves crashing, then retreating -- leaving behind a wet and unpredictable habitat where only a few sturdy creatures can survive.

Rough-girdled chitons (*Ceratozona squalida*) roam over wet rocks at night and on dark days. They have eight overlapping plates each of which contains microscopic pores that contain sense organs. They're hard to see because their greenish-brown coloring blends with their mucky surroundings. Their species name, *squalida*, is a Latin word meaning dirty.

I noticed zig-zag designs on rocks for years before I realized these tracks were made by limpets grazing on algae. **Cayenne keyhole limpets** (*Diodora cayenensis*) feed by moving slowly up and down rocks as their teeth scrape algae, leaving zipper tracks behind them. Covered with algae themselves, they're difficult to spot and they certainly don't move fast enough to attract my attention.

A pair of tentacles with eyes guides the limpets' grazing. They always return to their home base which is a small shallow depression in its rock. The "keyhole" serves as an outlet for water that enters their shell. The water supplies oxygen, then carries away waste products.

Both chitons and limpets clamp tightly onto rocks and are dependent on tides to carry nutrients and moisture to their rock habitat.

Checkered nirites (*Nirita tessellata*) are abundant on wet, rocky areas. I find these small black and white snails nestled in crevices and holes on slimy rocks. Their distinctive designs make them easy to identify.

Another rock resident is the **periwinkle** (*Tectarius muricatus*), another slow-moving mollusk found on rocks.

"The carbon in our tissues, the iron in our blood, and the calcium in our bones are all components of a universe that burst forth perhaps fifteen billion years ago. We are not just protoplasm; we are stardust. All our dust collectively brightens the planet we live on."
- Shirley Jones in *The Mind of God and Other Musings*

The Nature of Florida's Ocean Life

 Night Creatures

Florida has two tides daily, each reaching its highest point every 12 hours and 25 minutes. That's a 25-hour cycle which worms, mollusks and lots of other creatures (including me) guide their lives by. This cycle is much closer to my own body's rhythm than the 24-hour routine I follow so I can fit in with the un-natural world. Thousands of other night people complain about not fitting in -- but what's a body to do?

At night, raccoons, skunks, bats, rats, crabs, snakes, no-see-ums, mosquitoes, sand fleas and sea roaches become active scavengers, prowling in the quiet dark hours. Why do the least popular animals seem to be night creatures? Do they have bad reputations because they're nocturnal ... or did they become nocturnal because of their unpopularity?

Only in name do **sea roaches** (*Ligia exotica*) have any relationship to our more familiar cockroaches. They are isopods, which are more closely related to lobsters and shrimp. Secretive, they hide between rock crevices and usually come out only at night to scavenge along the rocks for stinky particles of seafood.

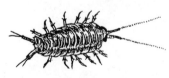

"The Sea Roach can drink water in the usual manner through its mouth, but prefers to drink through its anus. While clinging to a piling, the Sea Roach lowers its rear end into the water, enlarges its rectum, and then takes in water by reversing the direction in which fluids normally flow through the body."
- Winston Williams in *Seashells and Other Seashore Creatures*

Sand hoppers (*Orchestia platensis*) jump energetically around accumulated seaweed and ocean debris. Their scientific name *Orchestria* comes from the Greek word for dancer. And that's exactly what they are. They dance by the hundreds at my feet when I walk close to them, harmlessly hopping on and off my feet with a frantic energy that makes my own hyperactivity seem sluggish. These ⅛-inch creatures are also called beach fleas.

"For us believing physicists the distinction between past, present and future is only an illusion, even if a stubborn one."
- Albert Einstein

Algae

Some of the shells I find are so mutilated and deteriorated that I can't even identify them. The damage is caused by predators including sponges (which bore through the shell) and moon snails (whose acid secretion makes a neat little hole). But I was surprised to learn from Dr. David Cox of the Environmental Learning Center in Vero Beach that some of the shell's destruction is caused by algae. Marine fungi and algae eat through the calcium carbonate shells getting at the protein within the dead layers.

Algae are plants, but they don't have all the characteristics of our more familiar plants which have flowers and seeds. However, they are as varied in shape, color and size -- and as important -- as any of our land-based trees and flowers.

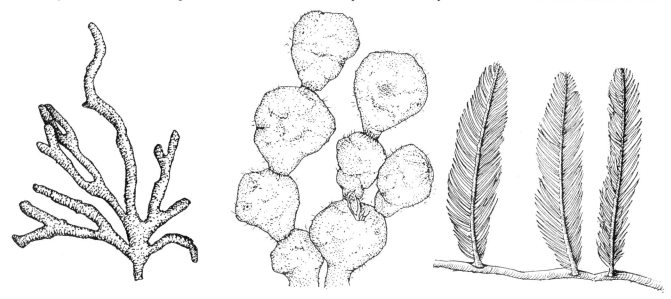

Above: yellowish green (*Codium decorticatum*), green (*Halimeda goreaui*), darker green (*Caulerpa sertularioides*)

Algae used to be one of the simplest forms of life; then their basic cells evolved, transforming them into the strangest and hardest organisms to identify. The species shown on this page are just a few that grow in Florida's coastal water.

Above: brown (*Turbinaria turbinaria*), red (*Heterosiphonia gibbesii* and *Amphiroa hancockii*)

"The one universal ever-operating law throughout has been the law of change. Nature never stands still and never duplicates herself. Life is always in the process of becoming something else."   - Laurence M. Gould

The Most Abundant Bird in the World

Shown below in flight (from left): **herring gull** (*Larus argentatus*), **common tern** (*Sterna hirundo*), **arctic tern** (*Sterna paradisaea*) and **Wilson's storm petrel** (*Oceanites oceanicus*).

The Wilson's storm petrel is probably the most abundant bird in the world as well as the smallest sea bird, at seven inches. Does their size allow them to survive more readily? Or is it because they're so remote from the rest of us?

If they're so abundant, why don't I see these petrels more often? Because they don't touch land except when breeding in the Antarctic.

Petrels travel *10,000 miles twice a year*, most of the time out of sight of land. They breed in the Antarctic islands, then head for the Gulf Stream, arriving along our coast around April.

Skimming over the ocean, they search for food; small fish, squids, crustaceans. Anglers notice the areas where petrels are diving, knowing that where there are small fish, larger fish are sure to follow.

During my night walks in the spring, I sometimes hear a soft peep-peep-peep by the shore. Even though I can't see the peeper, I'm sure it's a storm petrel feeding in the surf.

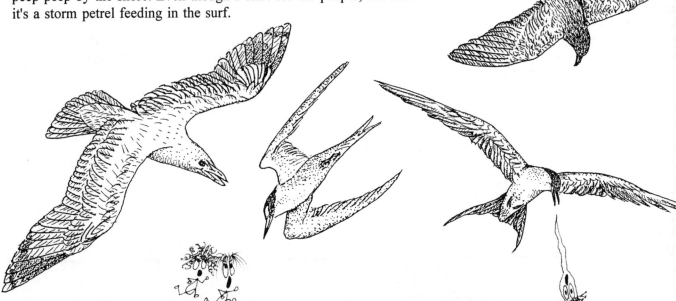

Red legs, red bills (tipped in black) and black caps make common terns easy to identify. They are typically spring and fall migrants to Florida, although some of them linger through the winter.

Common terns hover in the air, targeting small fish, then dive straight down into the water to their prey. They are often seen diving with other sea birds such as Arctic terns.

Arctic terns are similar to common terns, but some of the visible differences are their bills (not tipped in black) and they are slightly larger and have a longer tail. They migrate from the Arctic to the Antarctic where they molt, then return to the Arctic to nest.

"I believe that the scientist is trying to express absolute truth and the artist absolute beauty, so that I find in science and art, and in an attempt to lead a good life, all the religion that I want." — J. B. S. Haldane, British geneticist

Ocean Birds

Ocean birds rely on the sea's giant bowl of fish soup for all their food and liquid. Of the 8,600 species of birds, only a few are truly birds that wholly depend on the ocean. Migratory birds such as albatross and petrels are equipped with a salt gland that excretes excess salt to prevent dehydration. Most of our more familiar gulls don't have this gland and wouldn't be able to survive solely on the ocean.

Shown flying below: **greater black-backed gull** (*Larus marinus*), **magnificent frigatebird** (*Fregata magnificens*) and **northern gannet** (*Morus bassanus*).

The **man-o-war bird** (magnificent frigatebird) has a wingspan of *more than two yards*. Large colonies live in South Florida, but I often see one or two soaring high above the shore in Central Florida. They glide for miles without flapping their wings, and they can even swoop into the ocean to pick up food without breaking flight.

They eat fish, squid, jellyfish, gull chicks and sea turtle hatchlings.

We rarely see them perched because their short legs and broad wings make it difficult for them to take off unless they're on a tree top or rock.

During courtship, male frigatebirds inflate their bright red pouches to attract females. A male frigatebird will puff himself, clatter his bill and flap his wings wildly to get the attention of females.

Anglers also pay attention to **gannets** circling over the ocean; this is a signal to them that schools of small fish are close. Gannets are about the size of a big duck, but with a wing span up to six feet. Mature gannets are white with creamy yellow heads. In the winter I usually find several dead gannets washed onto the beach -- casualties of storms or maybe starvation. After Hurricane Andrew, hundreds of gannets died along Florida's coast from exhaustion and losing their familiar resting grounds.

"People are always blaming their circumstances for what they are. I don't believe in circumstances. The people who get on in this world are the people who get up and look for the circumstances they want and if they can't find them, they make them."
- George Bernard Shaw

A Little Whale Stuff

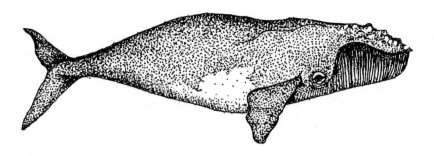

Each winter, **Atlantic right whales** (*Eubalaena glacialis*), returning from the north, can be seen about one to nine miles off Florida's east coast. Calving mothers come to our warmer coastal waters from November through March, sometimes bringing juveniles from previous calvings with them. When the northern waters warm up in the spring, the mothers return with their calves to the plankton-rich Cape Cod area. Then in July, they mysteriously disappear for awhile before re-appearing off the coast of Nova Scotia to meet other single whales for mating. Females mate with up to 30 different males each season.

The right whale got its name because it was the "right" whale to hunt in the days of serious whaling: the right whale was slow-moving and floated after being killed, making it easy for whale hunters to retrieve.

An unusual feature of right whales is the formation of *callosities*, thick skin patches which are visible even on newborns. Like fingerprints, each whale can be identified by their different patches. They occur on whales in all the same areas that humans have hair -- on the head, over their eyes (eyebrows), around their blowholes (mustaches) and chins (beards).

Whale Lice

Just because a mammal is big and lives in the ocean, doesn't make it exempt from life's little irritations such as lice. Lice, such as *Cyamus ovalis* and *Cyamus cigracilia*, stay with a particular whale for its entire life. Whales' skin is very thin for all their size, making for easy gripping for these lousy hitch hikers. The whale louse shown here is *Paracyamus boopis*.

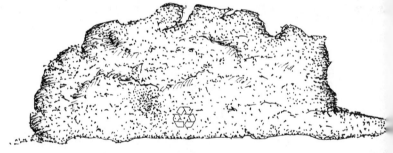

Ambergris

Periodically beachcombers come across waxy grayish lumps, but most people, not realizing its value, walk right past without a second glance. But this bland looking stuff is so extraordinary that poetry, songs and books have been written about it for centuries.

Herman Melville wrote about it in *Moby Dick*. And a few beachcombers in the past have become wealthy from it. The Chinese call it *lung yen*, "dragon's spittle" and nobles used it as an aphrodisiac. King Charles II of France's favorite dish was eggs and ambergris. In the days of great African kingdoms, ambergris, as valuable as gold and ivory, was traded through Arab merchants. *What is it?* Turn the page.

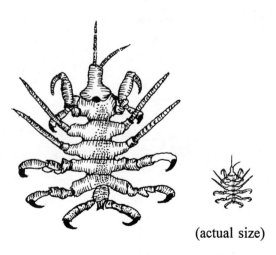

(actual size)

"I knew a marine biologist who was unable to talk to a molecular biologist because of the specialized nature of their disciplines. So, we have a situation where people know more and more about less and less."
— Roger von Oech *in A Whack on the Side of The Head*

Whale Gut Treasure: Ambergris

What *is* **ambergris** and what makes it so special? It's a substance formed in the intestines of **sperm whales** (*Physeter catadon*) to protect their insides from being punctured by squid beaks. Squids, a favorite meal for whales, have sharp indigestible beaks that accumulate in the whale's intestine. Ambergris is a protective coating of spongy material, made mostly of cholesterol, encasing the indigestible beaks and any other damaging remains in their colon.

Periodically the whale discharges the whole mass. Globs weighing up to a thousand pounds wash onto the beach.

> Jacques Cousteau writes about a mass of ambergris found in a whale's intestine, weighing close to a *thousand pounds*: "... and what an intestine it is ... in a whale 55 feet long, the intestine measures a thousand feet."

My associate, Glenn Bertiaux of Indian Harbour Beach donated a large chunk of ambergris to me and The Drifters.* He retrieved the musky-smelling ambergris from rocks on Ambergris Cay, in the Bahamas. The island is so named because ambergris used to be abundant there while the sperm whale population was still plentiful.

What makes ambergris so valuable? Foreign perfume industries use ambergris in their products to hold the fragrance long after the bottle has been opened. (Selling or trading it is no longer legal in the United States.)

More recently, medical studies are showing that ambein, a major constituent of ambergris, has properties used in anti-inflammatory drugs and other medicines.

Were sperm whales hunted solely for this ambergris? No. In the head of each sperm whale is *three to four tons* of *spermaceti* (hence, their name) which is a highly valuable substance used as a lubricant in fine machinery.

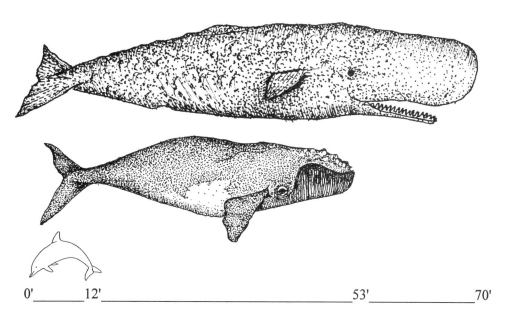

Approximate size comparison: dolphin (12'), right whale (53'), sperm whale (70')

"Next to the opals are open clamshells ... in each one, pearls have been formed by the coating of stray grit with a smooth gemlike luster. Sperm whales produce ambergris in a similar way to hide jagged annoyances (squid beaks). Both result in great beauty, and I love looking at them because they are a zen-like reminder that there are many ways to deal with irritations." - Diane Ackerman in "Slices of Life"

---

*The Drifters are an international group of beachcombers, naturalists and scientists who cherish items such as ambergris and sea beans. For information about our international newsletter *The Drifting Seed* and the Annual Sea-Bean Symposium, send a stamped, self-addressed envelope to The Drifting Seed, PO Box 510366, Melbourne Beach, FL 32951.

The Great Eel Mystery Solved: Migration

Eels are one of the most abundant and least understood creatures in our waterways. Freshwater eels in America and Europe have been marketed and eaten for centuries, but until recently most people thought of eels as snakes. Unlike other fish (which have well developed scales), eels have tiny scales which are embedded deeply within the skin.

Until this century, no one knew much about their lives, particularly how they reproduced. Young eels had never been seen ... nor had ovaries or testes been found in mature eels. How did they reproduce? Where did they spawn? Where were they born? Why didn't anyone ever see baby eels? And how do researchers study these nocturnal creatures which seem to disappear for part of the year?

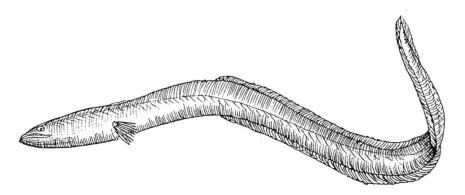

Eelbusters

"Eelbusters" are a group of researchers who study eels in the worst conditions -- under rocks in the dark and cold of night. Because of the eelbusters' dedication, we now have some answers about eel migration.

Meeting in the Middle

Freshwater **American eels** (*Anguilla rostrata*) leave Florida through inlets to travel east to the Sargasso Sea to spawn. About the same time, European eels (*Anguilla anguilla*) leave Europe to congregate in the Sargasso Sea too. (Although American eels and European eels look similar, a few traits distinguish them: American eels are larger and have fewer vertebrae.)

Before they begin their migration, adult eels develop an extra layer of fat to help them endure the long voyage. They stop eating completely while migrating, much to the disappointment of professional eelers who can't lure them into their traps even when baited with their favorite food.

By the time they reach deep water, the eels have developed huge eyes to let them see clearly in the ocean depths, and their skin has changed from murky green to shiny silver as camouflage in the shimmering ocean.

The king of eelers, the late George Robberacht, ran a million dollar business shipping eels to Europe. With large eels slithering over his arm he said, "American eels are harmless. Lots of people think they're slimy, but to me they feel like cool velvet."

Eel Life

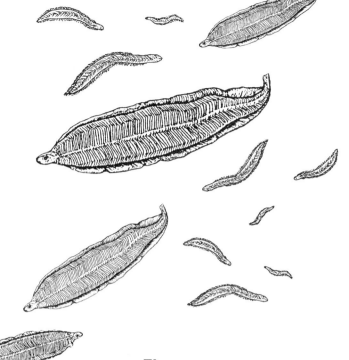

Eel spawning is still partly mysterious -- to see eels mating in deep water in the middle of the Atlantic underneath acres of floating seaweed isn't an easy accomplishment. But sailors from centuries past and researchers today have reported seeing millions of larvae floating with the seaweed.

The peculiar biology of eel larvae is that they look nothing like mature eels. Instead, they are transparent, flat and leaf-shaped, as shown to the right. (These are greatly magnified.)

Eel larvae are called *leptocephali*, from *lepto* meaning 'thin' and *cephali* meaning 'heads.' Anyone seeing these clear pin-heads would never suspect they become fat snake-like fish.

Ed Perry, an environmental park ranger at Sebastian Inlet and devoted inlet-angler says, "Larvae of ladyfish, tarpon and bonefish look almost exactly like the larvae of eels. The forked tail on the ladyfish larvae is hard to see but it's one of the only ways I can tell the difference. I'm amazed at how much they change when they mature."

Elvers

Millions of these larvae float around in the seaweed until they become *elvers*, a stage between larvae and adult, which still don't look like adult eels. (Shown left.) The young eels' bodies are so clear that their organs show through their skin as pink globs. Their eyes look like pairs of poppy seeds drifting through the water.

During this stage they travel with the Sargasso Sea, maturing until they're ready to return to their ancestral homes.

After one year of maturing, the American eels return to the freshwater where their parents lived. After three years of maturing, the European eels return to the rivers of Europe. Why do European eels take so much longer to return to freshwater?

And how do the eels know where to go? Why don't they ever get mixed up and swim to the wrong continent? After all, their parents died after spawning, so the elvers have no eel relatives to guide them home.

One theory about the strange migration of eels is that their origin was *Atlantis* thousands of years ago:

"Are they equipped with some built-in mechanism of recall that tells them the Sargasso Sea had been closed off in the east by some nearby land whose rivers eels once lived? The disappearance of such a land mass probably caused them to swim on, helped by the Gulf Stream, until they found the next-nearest land -- Europe. The disproportionate distance between the gathering place of these separate species and their respective habitats is otherwise quite difficult to explain."   - Don Groves in *The Ocean*

The Nature of Florida's Ocean Life

REEFS

"For everyone, reefs are daunting places. They support more kinds of life than any other area in the sea and they themselves are built by that life. Here, like no other place, different branches of science tend to merge. What are you to make of an animal or a plant which makes a rock? Is it a problem for biology or geology? Their complexity is such that they are visual chaos."
- Charles R. C. Sheppard in *A Natural History of the Coral Reef*

Sunken Ships

To think that human-made debris in Florida's ocean isn't *really* a part of our natural history is as naive as thinking that Disney World isn't *really* a part of Florida's geography. Sunken ships, dumped refrigerators and missile pieces are the foundation of many of Florida's reef habitats.

Several reefs created accidentally or intentionally by sunken ships attract divers from all over the world. In addition to Mel Fisher's well publicized treasure discoveries, Florida has pages of history still turning on the ocean's floor.

> "Florida is one of the best places in the world for such diving since the ocean surrounds it on three sides, and reefs, sandbars, and shallow channels have wreaked havoc on unsuspecting ships. Hurricanes, Confederate mines, enemy submarines, lightning, and dynamite have all taken a heavy toll on all kinds of vessels for almost 500 years."
> - Kevin McCarthy in *Thirty Florida Shipwrecks*

After Kip Wagner found a Spanish coin near Sebastian Inlet in 1964, he formed Real Eight, Incorporated, a treasure salvaging company. With Mel Fisher's Treasure Salvors, Incorporated, they began to search for Spanish Fleet ships. One of the salvagers' most significant discoveries occurred 40 miles off Key West when they found the *Atocha*, a seventeenth century Spanish ship that sank after running into a hurricane. Tons of gold and silver worth more than $100 million were recovered from this wreck.

Atocha's History

The *Atocha* left Havana in September, 1622. The ship's plan, after it was loaded with its valuable cargo from South America, was to sail from Cuba to the Florida Keys, then ride the Gulf Stream to Cape Canaveral. From there, the ship was to sail to Europe across the Atlantic.

However, the *Atocha* sank, killing hundreds of crew and passengers. Coins, gold, jewelry, dishes, books, equipment and furniture scattered over the ocean floor. News of the sinking spread to officials in Cuba who sent salvage crews to the site. Some of the treasure was recovered, but more hurricanes made it impossible to continue. So there it stayed until former chicken farmer Mel Fisher and his salvage team freed it from its centuries-old resting place.

> "They were ruined by their determination to accumulate as much wealth as they possibly could. This lust for gold did not enrich them; it destroyed them and the economy of their country, and compromised, for several centuries, the quality of human life in Spain."
> - Jacques Cousteau in *Diving for Sunken Treasures*

Sunken Treasures

Florida's waters have accumulated a couple thousand documented ship wrecks. Most are from twentieth century wrecks, but many sixteenth and seventeenth century ships still lie on the bottom, disguised as coral reefs.

> "If things had turned out differently, if we had found the treasure that we sought, it would have been the first time in my life that I gained material wealth from the sea; and that fact would have changed something in my relationship with the sea. I think that as a result, I would have been the loser rather than the winner..."
> - Jacques Cousteau in *Diving for Sunken Treasure*

Shown to the upper right, a gold whistle hanging from a 12-foot chain was found by Kip Wagner in the mid 1960s off the coast of Cape Canaveral. It was worn as a badge of rank by Captain General Juan Esteban Ubilla of the flagship of the 1715 Spanish fleet. The ship sank July 27, 1715 after a hurricane tossed it around like a toy. Nine other ships in the fleet also went down.

Progress

Treasures are still being found on Cape Canaveral beach -- nails, coins, buckles, pottery, chest lids -- reminders that less than 300 years ago, travel from Europe to America was dangerous and unpredictable.

From this same site in the Atlantic where ships crashed only three centuries ago, we now shoot missiles that splash into the Pacific Ocean on the other side of the world, minutes later.

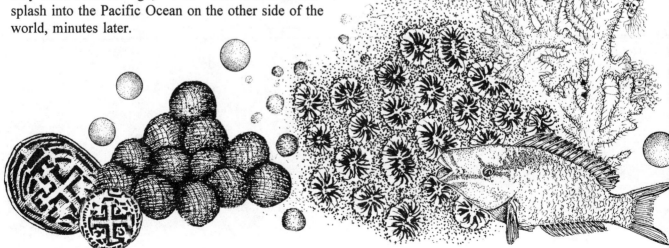

> "So, coral has covered the wreck. But the coral, in turn, has been eaten by parrot-fish; and the parrot-fish have excreted the coral in the form of sand. This sand falls to the bottom ... the study of the evolution of coral over a period of three centuries is challenging, but also discouraging; for it is likely that a great many sunken ships have disappeared forever beneath the rising ocean floor and will never be found."
> - Frédéric Dumas in *Diving for Sunken Treasures*

Human-Made Reefs

Visible at low tide from the shore at Vero Beach is a huge black mound, known as the 'Boiler Wreck,' jutting from the ocean. In April, 1884, the 300-foot English steamship *Breconshire* grounded just offshore.

After Captain Robert Taylor sent a telegram from Sebastian (then known as Quay) to inform England about his "total loss," the ship was stripped first by the crew, then salvagers, then locals. Most of the indoor plumbing in Indian River County back then came from the *Breconshire*. All that is visible more than 100 years later, is its huge black boiler. It sits about 300 yards offshore, an easy wreck for divers to see fish, lobsters, urchins and history.

In addition to the many accidental wrecks, other sunken ships were the result of pirates taking over and abandoning transport ships. After crew members were killed or captured, the pirates either took the ships or unloaded the treasures onto their own ships.

An artificial reef was created in 1985 by sinking a Rolls Royce in 90 feet of water just south of Palm Beach Inlet. A group of private citizens formed a committee to sink the Rolls Royce in response to Broward County's artificial reef which was created by a sunken ship named *Mercedes*.

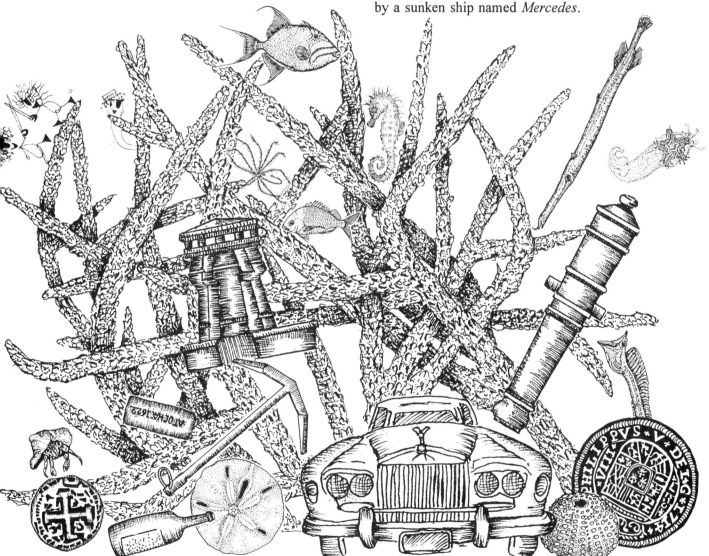

"Going to a junk yard is a sobering experience. There you can see the ultimate destination of almost everything we desire."
- Roger von Oech in *A Whack on the Side of a Head*

Coral Reefs

Reefs are busy condominiums of cleaners, workers, food providers, nurses, builders, and managers. Small fish keep the system clean by picking up particles; urchins graze on algae that could smother the coral; octopodes overturn rubble and rocks to stir out small creatures which in turn feed crustaceans and fish; sea slugs pick at sponges and anemones, cleaning them from marine debris that could choke the reef community.

Corals, sponges, tunicates, sea squirts, weeds, sea whips ... which is which and how do you tell the difference? What makes one a plant and another an animal? Plants get their energy for survival from the sun; animals need to eat plants or others other animals to survive. Coral animals (polyps) eat plankton (animal *and* plant).

"Biologists classify corals as animals, because they're obliged to call them *something*. But living coral tissue is actually composed of an extraordinarily tight partnership between animals related to sea anemones and certain single-celled marine plants. Their relationship, called symbiosis (literally meaning 'living together'), makes possible a quantum leap in ecological efficiency -- a way for both plant and animal partners to have their cake and eat it too."
- Joseph S. Levine in *The Coral Reef at Night*

Corals build their structure from calcium carbonate which remains even after the animal dies. Corals coexist with algae living within their tissues. These algae convert sunlight into energy, thus energizing the polyps to build their limestone (calcium carbonate) housing.

Millions and millions of these structures build on top of each other to create a coral system. Each type of coral has a different way of manufacturing its homes -- hence, the various designs and sizes.

Different personalities of corals are displayed in the reef system, ranging from passive to aggressive. However, witnessing these personalities is difficult, because everything happens in slow motion. Some corals will attack other invading corals with toxins. These corals have special sweeper tentacles which are longer and narrower than normal tentacles, and are loaded with stronger stinging cells.

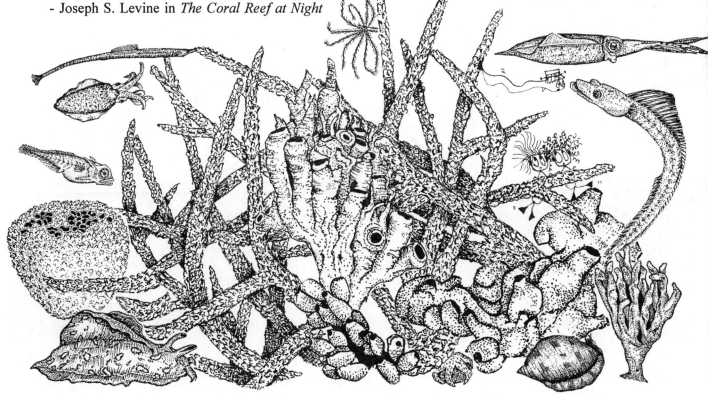

"All look frozen in still life. But we look on it with our sense of time, not theirs. In fact, a war is going on where species compete in many ways for survival. To us, it is slow motion, where a lightning strike by an aggressor may take many months! But it is none the less vigorous for all that. For the coral polyps there is a fight for survival as literally as there is on an African savannah, although with the difference that the participants cannot flee but must remain rooted to the spot of their initial attachment."
C. R. C. Sheppard in *Natural History of the Coral Reef*

Corals: Not Just Another Pretty Face

I used to think coral was some kind of lifeless rock that shell shops displayed -- pretty of course, but without much character. However, coral is actually the housing for millions of tiny animals who have a full and active social life. Their tentacles poke out at night, waving through the water in search of food. They reach into the sea soup, "shooting" at food that drifts by with microscopic poison cells.

As shown to the right, **stinging coral** (*Millepora alcicornis*) appears to be covered in stubble. Actually, their polyps are so thin they look like fine hair. Touching this coral will cause severe burning because the polyps contain "poison darts" that inject toxins.

Star coral (*Dichocoenia stokesii*) is one of our most familiar corals. I find pieces of it washed up on the beach, mixed in the wrack with other sea matter. Strong storms sometimes break off chunks, and because this coral is lighter than other corals, it easily drifts onto shore. Star coral with a close-up view is shown to the far right.

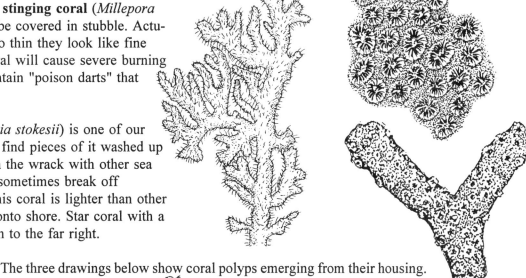

The three drawings below show coral polyps emerging from their housing.

Corals spawn predictably after a full moon in August. Each coral polyp develops sperm and eggs during the breeding season. Then clouds of sperm are released into the water, float to the surface and pair with receiving eggs. Currents encourage this mating game by keeping the water stirred. Shown right are coral *gametes*, bundles of eggs and sperm.

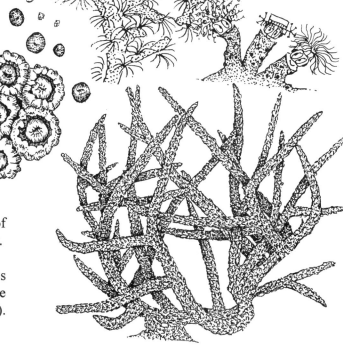

Staghorn coral (*Acropora cervicornis*) grows in colonies of forest-like structures in the reef system, as shown to the right.

Elkhorn coral (*Acropora palmata*) has tree-like structures similar to staghorn coral, but the branches are more like those of a moose rather than an elk (elkhorns are fatter and flatter). They can grow to 10 feet and cover miles of ocean floor.

"The beauty of the natural world lies in the details, and most of those details are not the stuff of calendar art." - Natalie Angier

Tongues, Cameos and Blennies

Flamingo tongue shell (*Cyphoma gibbosum*) is known as the leopard of the sea because of its spotted shell. But the orange and black spots actually come from the animal's *mantle*, the fleshy body that comes out and covers the shell (which is a solid creamy yellow). The animal grazes on soft corals, munching on the polyps and leaving a polyp-free trail behind them.

Cameos

Carving gemstones has been an art since 400 BC. Artists eventually began engraving shells as another way to make use of natural resources. Cameo artists cut through layers of shell, exposing hidden colors to create portraits and scenes commemorating historic events.

Conchs, cowries and helmets are the shells most used to create cameos. Although Florida isn't famous for cameo art, our **queen helmets** (*Cassis madagascariensis*) and king helmets (*Cassis tuberosa*) are used because of their width, thickness and layers of color. The world's cameo center is in Torre Del Greco, a small Italian town, where the world's only cameo school offers classes in the ancient art of cameo engraving.

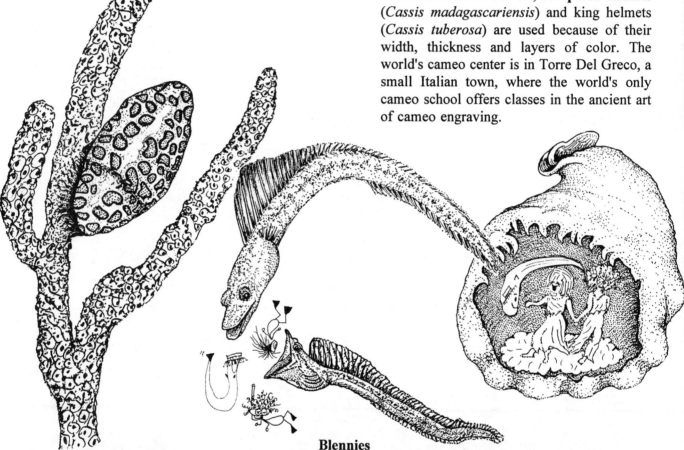

Blennies

Broken shells and pulverized coral accumulate on the ocean bottom. Divers see lots of little creatures, including fish, digging through these bits and pieces. Blennies are notorious for poking through sand, stirring up clouds as they look for food. In between eating, they scoot backward into holes, leaving only their heads sticking out.

Like other blennies, the **yellowface pike blennies** (*Chaenopsis limbaughi*), shown above, are comical for divers to watch. Only a few inches long, they look like little kids trying to act grown-up.

When threatened, this blenny acts like a cobra by slithering out of its hole and spreading its dorsal fin. It then flares its gills and weaves from side to side. Next, it strikes with mouth open wide, assuming a warrior's fierceness as much as a three-inch bug-eyed cartoon character can.

> "Shoot for the moon. Even if you miss it you will land among the stars."
> - Les Brown

Sponges

The variety of sponges living in the ocean surprises me. I used to think of sponges as those soft yellow blobs sold in Tarpon Springs and bath stores. But sponges range in size from less than an inch to over six feet; shapes vary from branched stalks to oval vases; colors cover the spectrum. (Often sponges lose their bright colors by the time they wash onto shore.) But one thing sponges all have in common: they all have holes.

Sponges belong to a group called *Porifera* which means 'pore bearers.' They are one of the most ancient animals, existing since about 580 million years ago.

How do sponges defend themselves? Their main defense is that they taste and smell bad to predators. And most sponges have microscopic hard spikes called *spicules* which are painful when swallowed. Touching a glass sponge will leave a diver's hand full of splinters that are hard to remove. But their best defense is persistence ... they just don't give up even when torn apart.

> "No matter how specialized sponge cells become they never lose their ability to lead an independent life ... If you grind up a sponge and squeeze it through a fine silk cloth, the cells that are still alive will regroup and grow into a new sponge ... After sponges made a beginning toward specialization nature seems to have gone back and started over, leaving sponges in an evolutionary blind alley."
> - William J. Cromie in *The Living World of the Sea*

Sponges can move when necessary: "Some cells actually broke free of the sponge and wandered about by themselves (looking like little lost amoebas) for awhile before reuniting with the 'mother ship.'" (Calhoun Bond, watching sponges crawl under a microscope.)

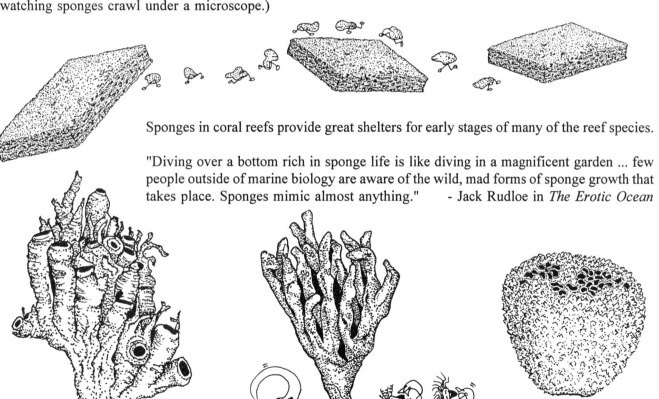

Sponges in coral reefs provide great shelters for early stages of many of the reef species.

"Diving over a bottom rich in sponge life is like diving in a magnificent garden ... few people outside of marine biology are aware of the wild, mad forms of sponge growth that takes place. Sponges mimic almost anything."   - Jack Rudloe in *The Erotic Ocean*

Some of the "mad forms of sponge growth" shown above are **yellow tube sponge** (*Aplysina fistularis*), **finger sponge** (*Haliclona oculata*) and **black ball sponge** (*Ircinia strobilina*) which spreads out like a huge boulder.

"I wonder how much deeper the ocean would be without sponges." (unidentified quote from the Internet)

The Nature of Florida's Ocean Life

Octopus and Squid Facts

Octopus, squid, snails and sea hares are all mollusks, the same group (phylum) as our land slugs and snails.

What's the difference between an octopus and a squid? An octopus has a bag-like body with eight arms, each with a double row of suckers; a squid has eight arms and an additional two tentacles. The two tentacles are used to capture prey. Both have ink sacs but an octopus' ink is black, while a squid's ink is bluish-black.

Female octopodes attach their eggs in strings to the roof of their holes. Embryos develop inside the eggs, feeding off a yolk sac. The OctoMom, as octopus expert James Wood calls her, guards and cleans the eggs with the tips of her arms or by squirting water over them. Eggs hatch into tiny versions of the adult, about 1/4 inch long. Octopus eggs are shown to the right.

Squid eggs are shown to the left.
"One of my best and favorite finds was a blob of squid eggs. It was still viable when I grabbed it and took it home. They were way, way cool. With the naked eye, you could see the embryos moving about in their capsules. And they already had developing eye stalks. Wow."   - Cathy Yow

An elaborate courtship ritual with graceful dancing and teasing lasts for several days among squid; sometimes females die from exhaustion or are eaten by the males. If they survive through the vigorous mating, they produce transparent, whitish egg sacs, draped over rocks and seaweeds.

Squid can range in size from thumb-sized to 50-feet giants. About four species can be commonly found in Florida's reefs.

One of them, the **Atlantic reef squid** (*Sepioteuthis sepioidea*) is also known as the popeyed squid because of its large eyes. They swim by rippling their fins. In the reefs they stay in schools of about a half dozen in the back area, but come out to investigate activity.

"The Atlantic reef squid, because of its gregarious and fearless habits, provides much joy to snorkelers and divers. Small schools of this wonderful animal appear suddenly in the corner of your eye as you swim over lagoon or reef. You are startled, not so much by their sudden appearance as by the realization that they are staring at you with as much interest as you are exhibiting toward them. Approach slowly and they will hold their ground, staring all the more intently. When you are about 5 feet away they will swim slowly backward in unison, with their fins rippling and their tentacles bunched up. Swim faster and they will match your velocity, keeping the same precise distance. Stop and they stop too. This game can go on for 15 minutes or more, until one of the parties tires. "
- Eugene Kaplan in *Southeastern and Caribbean Seashores*

"I never think of the future. It comes soon enough." - Albert Einstein

Octopodes

What's the correct plural for octopus? James B. Wood, professor and cephalopod expert, says, "While 'octopi' and 'octopuses' are both sufficiently common to be acceptable, 'octopuses' is three times more common. The purists' favorite, 'octopodes' [the correct one] is virtually never used."

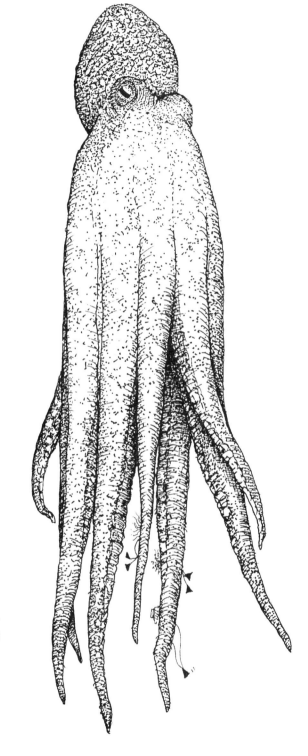

The **common reef octopus** (*Octopus briareus*), although about a foot and a half, can squeeze its bulk through tiny crevices. Without the hard shell of their mollusk ancestors, they are flexible and mobile, displaying a grace that contrasts with their lumpy bodies.

Can octopodes walk out of water? James Wood says, "Yes, they can and do voluntarily leave the water to go exploring, hunting, etc. I've seen *octopus briareus* do this in the Florida Keys ... I'm not sure how long they can stay out of the water though (they do have gills after all) - certainly long enough for a midnight snack."

Octopodes do most of their hunting at night, swimming or crawling along the ocean floor looking for crabs, lobster and mollusks. They pile the remains of their meals outside their hole; divers use this as a signal of an inhabited octopus hole.

Using their arms to grip prey with the suckers, they inflict the victim with a nerve poison. To eat something as hard as a lobster, the octopus punctures the shell with its tongue which is covered with sharp teeth.

Octopodes can change color and texture rapidly. The color change is a reaction of their pigment cells (chromatophores). Each cell may be one of several colors -- red, yellow, blue, black. When the cells expand or contract (by using certain tiny muscles), their color is either exposed or hidden. Octopodes change color so fast because the chromatophores are under neural control -- the same as when we decide to smile or wink.

"When an octopus is angry, it often changes color. If, for example, you surprise an octopus on the bottom and touch it near the head, it turns white, and then brick-red which is the color of discontent. It takes about two seconds for the entire body to change color. Then the animal holds its arms over its head, like a protective helmet, and runs away in that position."
    - Jacques Cousteau

"Many scientists nowadays spend too much time behind laboratory walls while they explore the ecology of computers ... Even worse, they become entrapped by bureaucracy. New Big Science demands the diversion of extraordinary amounts of career time to political strategies and grantsmanship designed to pry money from funding agencies. One wonders if potential Newtons, Darwins, and Einsteins are being quietly smothered by the bureaucratic octopus. We are losing the ability to stand and stare." - John L. Culliney in *Forests of the Sea*.

Strange Relatives

Sea stars, urchins, sand dollars and sea cucumbers all belong to a group called *echinoderms* which means "spiny skin." They all have a hard skeleton called a *test*; sand dollars and sea urchins are examples of these tests that we commonly find on Florida's beaches.

Feather stars are the oldest living group of starfish. Fossils of these stars show they haven't changed for 400 million years.

Feather stars have at least two (and up to 200) feathery arms extending from each of their five body segments. These arms wave in the water, catching food particles as they drift by. Divers know how fragile these beautiful creatures are, so they're particularly careful not to accidentally break off any of the arms.

Feather stars generally stay anchored to sea whips or corals (or to themselves). But some, like the **swimming crinoid** (*Analcidometra armata*), can swim through water by waving its feathers gracefully up and down.

Florida oldtimers say that **sea urchins** give us clues about weather: they bury themselves in mud just before storms. I don't know about that, but I do know that sea urchin roe, *uni*, is a delicacy at sushi bars.

Sand dollars are simply flat urchins. Short bristles are constantly moving as they roam. This kicks up sand around them, camouflaging their sand-colored bodies. At the same time, particles of food are stirred out of the sand for them to eat. When they sense danger, these flat urchins can disappear below the sand within seconds by "digging" down with their bristles and tube feet.

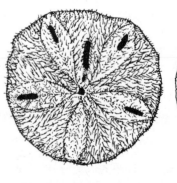

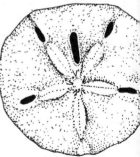

Sea stars are able to flip themselves over after landing upside down from a tumble in the surf.

The **five-holed keyhole urchin** (*Mellita quinquiesperforata*) shown here with and without its bristles. The bristle-less one is typical of how we find them washed up on the beach.

"I don't know what your destiny will be, but one thing I do know: The only ones among you who will be really happy are those who will have sought and found how to serve."   - Albert Schweitzer

Reef Creatures

My beachcombing buddies and I used to call sea cucumbers *izzies* -- every time we'd find one, we'd say "Izzy dead or izzy alive?" Izzies were the slowest creatures we had ever seen -- we'd check on them throughout the day, nudging them with our toes to see if they'd react (they didn't) -- but at the end of the day, they'd be in a different spot.

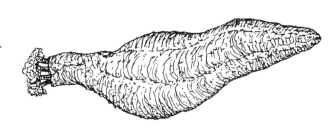

The **sea cucumber** (*Thyonella gemmata*) shown to the right has its tentacles extended to grab food from the water. Sea cucumbers also use their tentacles for reproduction. During spawning, male cucumbers extend and retract their tentacles to emit sperm into sea grasses. Females contract and expand their tentacles to release thousands of tiny blue-gray eggs that mingle in the grass with the sperm.

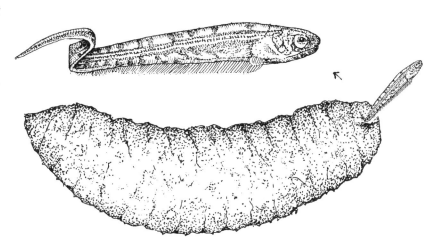

The larger **Agassiz's** sea cucumber (*Actinopyga agassizii*) grows to about a foot long and has an intimate relationship with a nearly transparent fish called **a pearl fish** (*Carapus bermudensis*). This fish stays inside the sea cucumber's gut, just inside the anus. Supplied with a flow of water going in and out of its home, the pearl fish only leaves its host at night.

Green morays (*Gymnothorax funebris*) are common in Florida's reefs, hiding under rocks and in crevices. Generally, they stay hidden during the day, coming out at night to feed on small fish, octopodes, crabs and mollusks.

Morays look mean with their huge mouths gaping all the time, but that's just how they breathe. The green moray is the largest of several species of morays in Florida and can grow to six feet. Their color comes from a yellowish mucus which covers their much darker skin.

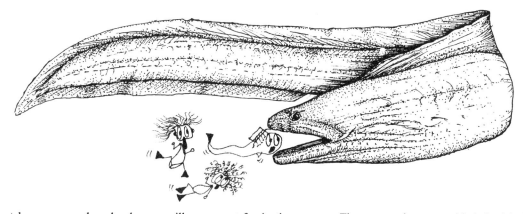

"If you watch a sea cucumber closely, you will see a most fascinating process. They sweep the water with their eight long, sticky, finely branched tentacles and capture bits of planktonic life, then they wipe the food off with two smaller tentacles, much as a child licks candy from his fingers."

- Jack Rudloe in *The Sea Brings Forth*

Lobsters

I'm familiar with two types of lobsters: the American lobster (*Homerus americanus*) which is the big-clawed one that restaurants serve; and **Florida's spiny lobster** (*Panulirus argas*). However, I didn't know until recently how many different lobster species existed in our southern ocean.

Don Bradley, a biologist from Melbourne Beach, discovered two species when he was a graduate student. One was named after him and the other after his grandfather: *Munidopsis bradleyi* and *Chirostylus blesseyi*. Don informed me that about 45 species of spiny lobster exist in our ocean. Who knew?

Our familiar spiny lobsters hide in crevices in the reefs by day and come out at night to feed. They have no pincers, but they have long antennae. When they move their antennae up and down, they produce a rattling or squeaking sound that warns other lobsters to stay out of their territory.

They use their tails to move backward, by "scooting" themselves along the sandy floor.

"How many species of organisms are there on Earth? The number could be close to 10 million or as high as 100 million ... It is a myth that scientists break out champagne when a new species is discovered ... We don't have time to describe more than a small fraction of those pouring in each year."

- Edward O. Wilson

Lobster Migration

Every autumn off Florida's coast, a remarkable migration of spiny lobsters takes place. They line up single file, in queues, the flexible antennae of each animal constantly touching the tail of the animal in front. This parade may include up to 65 lobsters, both male and female, moving from shallow water to deep water.

They gather for the great migration after a cold front. Since lobster populations aren't as great as earlier in the century, these migrations aren't as common as they used to be. But! When they march -- what a sight.

Why the Parade?

Pushing their way through water, a queue of 19 lobsters can reduce drag by about 65 percent over the drag each lobster would experience alone. Conserving energy is important for them to get from inshore to offshore for breeding in the spring. Another reason for the queue is safety in numbers.

Does the same lobster always lead? No, leaders change as the queues break into smaller lines, sometimes joined by other queues later and switching leaders again.

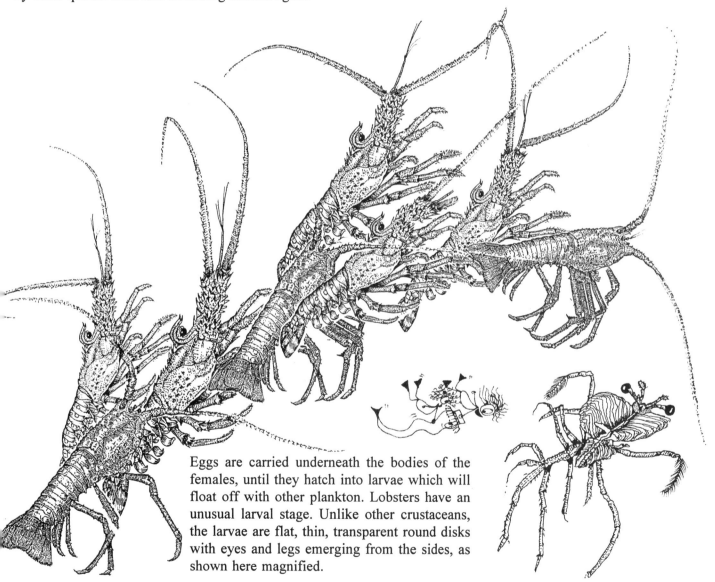

Eggs are carried underneath the bodies of the females, until they hatch into larvae which will float off with other plankton. Lobsters have an unusual larval stage. Unlike other crustaceans, the larvae are flat, thin, transparent round disks with eyes and legs emerging from the sides, as shown here magnified.

"The recent discovery that adult lobsters are strangely attracted to certain refined petroleum products, notably kerosene, is another danger signal. The animals can detect kerosene at fantastically low concentrations in sea water. The stuff so powerfully stimulates searching and feeding behavior that lobstermen have begun using kerosene-soaked bricks to bait their traps."
- John L. Culliney in *The Forests of the Sea*

The Nature of Florida's Ocean Life

Shrimp

When I think of shrimp, an image comes into my head of white, tasty meat dangling over the edge of a shrimp cocktail glass. So I was surprised to learn about the mean-spirited nature of a shrimp that looks and acts like a praying mantis (hence, the name). Two of my favorite nature writers describe **mantis shrimp** (*Squilla empusa*) like this:

"...when I think of sea lice [mantis shrimp], I think of a monstrous-looking shrimp creature that twists and flexes its jagged tail and cuts deep into any unprotected hand, drawing blood. I remember how I first picked one up and how it slashed my fingers and I bled profusely ... 'sea lice' is a good honest name for them."
    - Jack Rudloe in *The Sea Brings Forth*

"These claws snap out to strike rivals or seize prey in one of the fastest animal movements ever recorded: six milliseconds from start to bloody finish."
    - Jennifer Acherman in *Notes from the Shore*

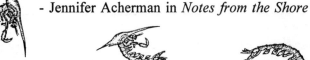

Even their ¼-inch babies look mean:

These aggressive mantis shrimp live in burrows and stalk the ocean floor, looking for prey to smash with their huge, hard-shelled knuckles. Appropriately named, mantis shrimp resemble praying mantis insects in savagery, aggressiveness and the way they hold their prey.

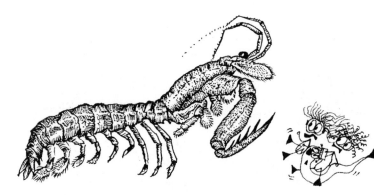

Their blows, like karate chops, hit with a force of a small caliber bullet. They can break an aquarium made of two layers of safety glass, and can shatter a thick-shelled clam with a single whack. They sneak up on crabs from behind and kill them with a swift hit, and then smash both of the crab's claws before devouring their meat. Yikes. Does anything redeem these shrimp? Yes, they're tasty, but who wants to catch these Saddams of the sea?

A Kinder, Gentler Shrimp

Pederson's cleaner shrimp (*Periclimenes pedersoni*), less than an inch long and transparent, stays within an anemone's tentacles, immune to the stings in exchange for its cleaning services. Their cleaning instinct is so strong that divers say if they hold their hands out, the shrimp will clean their fingers.

In addition to its cleaning skills, the shrimp's long antennae sway back and forth attracting small fish toward its host. What do the shrimp get in return for their duties? Beside getting protection from predators, they receive free meals as fish particles swirl near them from the anemone's scraps.

Shown here on a corkscrew anemone, this cleaner shrimp is so transparent, you can see rows of eggs inside the abdomen.

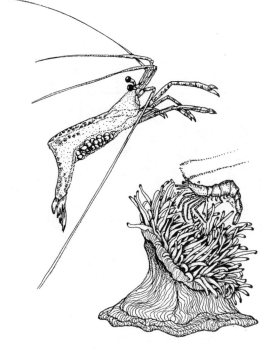

"There are some things that are so serious that you have to laugh at them."
    - Niels Bohr, physicist

Reef at Night

At night, the reef becomes noisy. Fish chirp, pop, grunt and wheeze, swish, click and groan. In World War II, sonar operators in the U.S. Navy "heard" suspicious snapping sounds. After they investigated, they discovered the noise came from a tiny snapping shrimp. The operators described the sea noise as "sizzling, like burning brush or frying bacon."

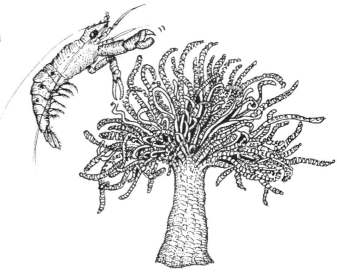

The sonar operators were hearing the **pistol shrimp** (*Alpheus armatus*). Not more than two inches long, this little crustacean makes a sharp pop, like a cap gun. Each of its front legs has a claw, one of which is grossly oversized. A "finger" on this large claw slams like a trigger against the opposite finger, creating a snap in the same way we snap our fingers. But our fingers contact with soft flesh, rather than a rock-hard shell. This snap is powerful enough to stun a small fish so the shrimp can grab it, and loud enough to scare off predators.

The shrimp's body is bright red with white spots to blend in with its host, the **corkscrew anemone** (*Bartholomea annulata*). The corkscrew (or ringed) anemone gets its name from its long droopy tentacles that are ringed with a light brown, giving the impression of being curly. These rings are actually "batteries" of stinging cells.

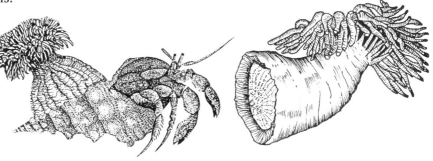

Some anemones can detach themselves and drift away from a predator when necessary.

The **tricolor anemone** (*Calliactis tricolor*) is one of the most common anemones found on the beach after storms and extreme high tides. I find them attached to empty shells and lumps of sponge.

In their normal habitat, the anemones attach themselves to shells occupied by hermit crabs. In this symbiotic relationship, the anemone gets to travel and pick up leftover scraps of food from the crab; the crab, in turn, gets protected by the anemone's stinging tentacles. Periodically the hermit needs to switch to a bigger home; when it switches, it takes along its companion anemone by tickling around their attached base, then positioning them onto their new shell.

Brooding anemones keep their little ones close. They are born within their parent's body and emerge through the mouth. The babies attach themselves to the base of the adult where they remain, protected, for their first few months.

"A day without sunshine is like, you know, night." - unidentified quote from the Internet

Reef Worms

Christmas tree worms (*Spirobranchus giganteus*), shown far right, live in hard tubes within living coral. Each has two spiraled crowns with a double-horned "plug" (like an operculum of a mollusk) used to close itself inside its tube. The tentacles wave through the ocean water to catch particles of food.

A **bearded fire worm** (*Hermodice carunculata*), shown right, is a slow-moving creature with short, white bristles poking out from red stalks. The flesh on the this worm's head is called a *caruncle* and is used to taste the water for food.

How does a slow-poke like this fire worm capture food? It preys on food that can't move away, like coral and anemone polyps.

Divers are careful not to touch the glasslike bristle hairs which easily detach on contact. These bristles cause severe pain that can last for several hours.

Feather duster worms (*Hydroides* spp.), shown below, immigrated to our waters from Europe by attaching themselves to ships' hulls.

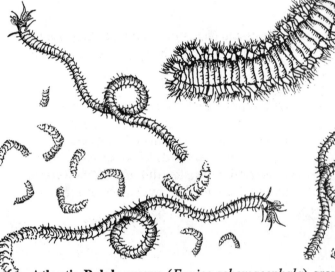

Atlantic Palolo worm (*Eunice schemacephala*) grow to about six inches long and live inside old coral, boring and breaking down reefs with their complex mouths (shown above).

They have gained fame from their unusual and predictable mass reproduction: a specialized section develops from the worm's rear. This section breaks off and swims to the surface where thousands of these strange rear pieces release sperm and eggs. (The rest of the body stays behind to grow a new rear.) Divers in the Dry Tortugas see the mass swarming by these rear end pieces, and describe the ocean water as turning into "thick, vermicelli-like soup." This occurs every year within three days of the moon's last quarter, between June 29 and July 28. The fertilized eggs develop quickly into microscopic creatures that eventually settle on the ocean bottom

If these tiny creatures are so strongly influenced by tides and lunar pulls, why should I be surprised when I feel the force of a full or new moon pulling on me?

"The science of life is a superb and dazzlingly lighted hall which may be reached only by passing through a long and ghastly kitchen."
- Claude Bernard, French physiologist

Fish

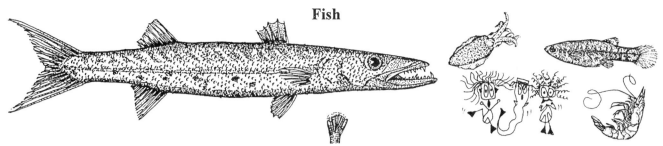

Great barracuda (*Sphyraena barracuda*), despite their savage reputation, rarely attack humans. They're attracted to bright and shiny objects, so a diver wearing jewelry might increase her chances of attack. Shaped like torpedoes, they blast toward other fish, squid and shrimp.

Barracuda may be tasty, but a toxin called *Ciguatera* can make them dangerous and poisonous to eat. This results from barracudas in the Atlantic eating lots of toxic prey which is abundant in our reef system. However, California barracudas have never been known to be poisonous -- and the Pacific reefs off California don't have the number of toxic species that we have in Florida.

Trumpetfish (*Aulostomus maculatus*) (right) look like the elongated plants and blades of sea grasses they use for camouflage. They align themselves vertically, head down and suck prey into their mouths by pouncing on them from this vertical position.

Queen triggerfish (*Balistes vetula*) can change color to blend with changes in their background, but the bright blue stripes on the head never change.

One of the queen triggerfish's hunting techniques is blowing a stream of water at an urchin until it flips over. This exposes its undersides which the triggerfish then attacks to get the soft meat inside.

Queen parrotfish (*Scarus vetula)* constantly browse on the coral, scraping off not only the polyps, but huge amounts of limestone which they poop out as clouds of white sand. Wherever parrotfish are plentiful, tons of pulverized limestone accumulate over acres of sea floor.

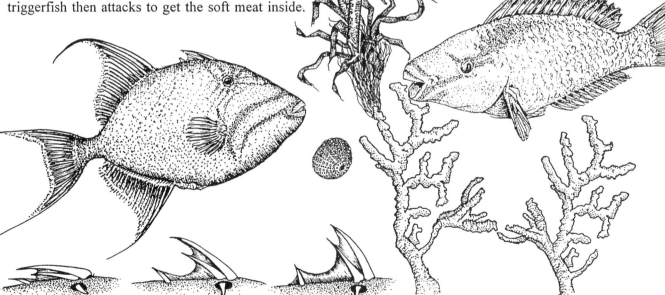

When not actively hunting, triggerfish anchor themselves inside crevices by *triggering* their dorsal fin to lock. A trigger mechanism at the base of the second spine (shown above) releases the first spine in order to lower the fin. When the trigger is activated, the fin straightens, thus anchoring it in place. No predator can pull the fish out once it's locked in a crevice.

"The zest, the fire, the savor of existence comes from something deeper, something spontaneous, native, and protoplasmic, which we can never outgrow or avoid, not should we wish to do so."   - Edmund W. Sinnott, U.S. biologist

The Nature of Florida's Ocean Life

GULF STREAM

"There is an odd flow of water in the Atlantic Ocean which has had a profound effect on human beings. This current of water has been called a 'crazy river' by some ... In the past it controlled the discovery of the Americas, and has been partly responsible for some of the more sordid episodes in history ... The Gulf Stream could control battles in the days of sail, and it can affect the tactics of submarine warfare today. One day the warmth of the ocean currents may prove to be a cause for argument."
-Thomas F. Gaskell in *The Gulf Stream*

World Travelers

Snake in the Sea

The Gulf Stream is a huge snake-like river of warm water slithering north from the tropics and moving parallel to the U. S. coast. As it flows along Florida's shore, it periodically curves in, but veers away when it gets just close enough to touch land. The Gulf Stream by Palm Beach is less than three miles offshore; by Melbourne Beach, it is 20 miles offshore.

From its beginning far south of Florida, the Gulf Stream displays an independence and determination: it forces its way between Cuba and Florida, squeezing its enormous bulk through the narrow Straits, then blasting out with the power of a firefighter's hose. Fifty miles wide, this powerful Stream keeps its identity for thousands of miles as it travels north. The Gulf Stream moves faster than any other current in the open sea.

Eventually, the warm Gulf Stream meets the icy Labrador current off the Grand Banks of Newfoundland, where notorious thick fog results from the union.

Continuing east to Europe, weaker and slower, the Gulf Stream splits, one part heading northeast toward the arctic, another part turning south toward Africa.

Looking at the Gulf Stream from an airplane, I've seen its deep blue water contrasting with the cloudy green of the Atlantic Ocean on either side. I try to imagine its tremendous power from my perspective -- and I can't even fathom the complex energy needed to transport tons of fragile eggs, larvae, seeds, spores, algae, weeds, logs and fish from exotic tropical islands to my beach in Central Florida.

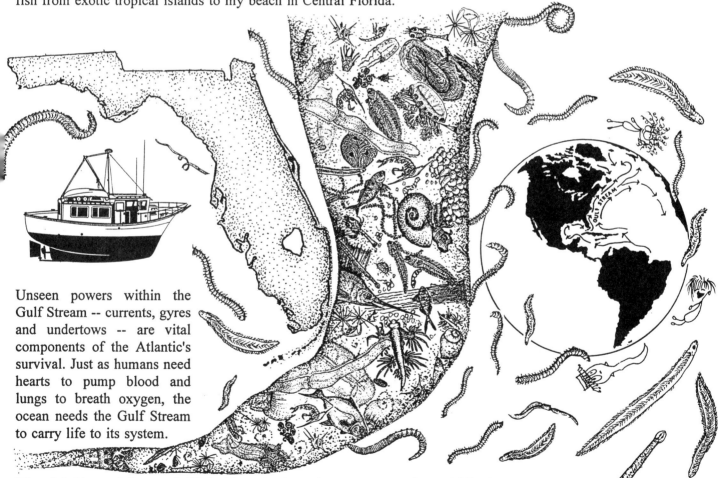

Unseen powers within the Gulf Stream -- currents, gyres and undertows -- are vital components of the Atlantic's survival. Just as humans need hearts to pump blood and lungs to breath oxygen, the ocean needs the Gulf Stream to carry life to its system.

"The whole history of discovery is filled with people who used erroneous assumptions and failed ideas as stepping stones to new ideas. Columbus thought he was finding a shorter route to India."
                -Roger von Oech in *A Whack on the Side of the Head*

The Gulf Stream: A Little History

What is the power controlling this huge underwater river? Is it a force leftover from billions of years ago when the Earth only had one huge continent? Is its strength a remnant of five million years ago when Panama was still underwater -- when currents flowed "naturally" since nothing separated the Atlantic from the Pacific?

When Panama rose from the sea, it created a wall between the two oceans. How did the free-flowing current react when it met newly formed land? Pushed away, the rejected current joined forces with other, already established currents, and -- voilá: the Gulf Stream ... with a power and speed surpassing all others in the world.

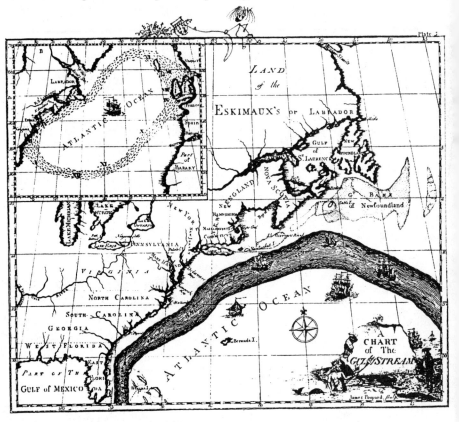

A Little History

When Columbus planned his journey across the Atlantic, he knew he would reach land: sea beans, bamboo stems and coconuts occasionally washed up on European shores. He figured, correctly, that the debris indicated both land to the west and a current traveling east.

By the early 1500s, Spanish ships had discovered the advantage of sailing with the good currents, using the Equatorial Current to cross the Atlantic, then riding back home on the Gulf Stream.

Benjamin Franklin was the first to put the Gulf Stream on a map, as shown to the right.

What was Franklin's interest in the Stream? When he was Postmaster General in 1770, he wondered why merchant ships from Europe returned two weeks faster than the mail ships. Merchants knew from whale hunters about the Stream's power and direction. Franklin used the whalers' information and, in combination with his own temperature measurements inside the Stream, he defined, for the first time, the Stream's boundaries.

In July of 1969, a history-making event occurred in the Gulf Stream: Professor Jacques Picard traveled *in* the Gulf Stream for one month in a 48-foot submersible, the *Ben Franklin*. This was the first time anyone had seen the Gulf Stream from the inside.

While Professor Picard was drifting in the Gulf Stream in his sub, Apollo 11 was on its way to the moon. When Astronaut Edwin Aldrin gazed at Earth through the spacecraft's window, he radioed to Earth, "Say, Houston, you suppose you can turn the Earth a little bit so that we can see a little bit more than *just water?*"

"Humans with their terrestrial bias, forget that the 'rivers' in the sea are no mere superficial features such as the Mississippi or Amazon, a few tens of meters deep. The Gulf Stream alone contains over a hundred times more flowing water than all the terrestrial rivers on earth. Off the southeastern United States, the current reaches more than a kilometer deep, filling a vast invisible water-valley whose fluid walls are up to eighty kilometers wide, and it touches regions that are still more remote to human experience than the moon."
- John L. Culliney in *The Forests of the Sea*

Sea Soup

Sea wasps (*Chiropsalmus quadrumanus*): "Daytona Beach suffered an outbreak of these crystal-clear, bell-shaped devils that moved quickly, quietly and unnoticed through the shallows where bathers were enjoying the surf. Suddenly cries of pain were heard and people fled the water. They never saw what hit them."
- Jack Rudloe in *The Erotic Ocean*

The **upsidedown jelly fish** (*Cassiopea frondosa*) has a frisbee-like body with frilly lettuce arms. It swims rightside up, but when it goes to the ocean bottom it hovers, then turns over, landing like a flying saucer on a cratered surface. Their toxic sting can leave a welt, so divers know to avoid touching them.

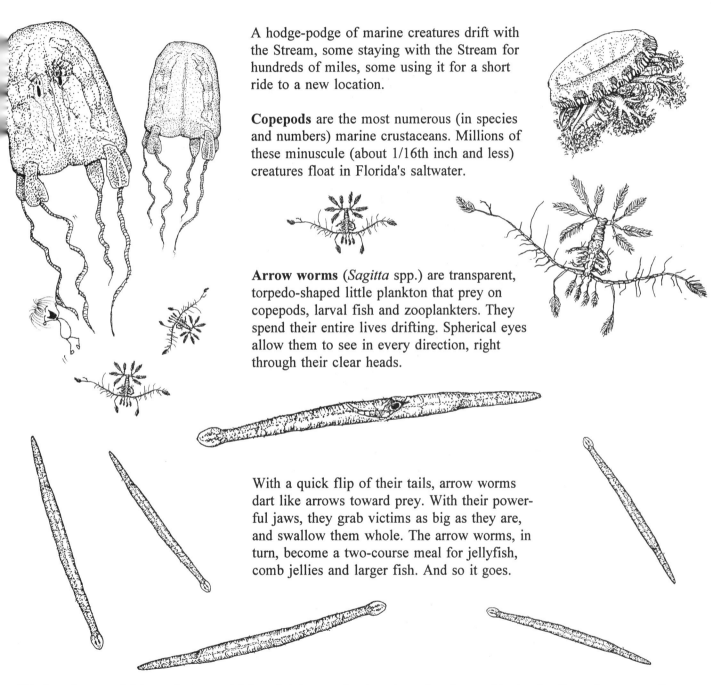

A hodge-podge of marine creatures drift with the Stream, some staying with the Stream for hundreds of miles, some using it for a short ride to a new location.

Copepods are the most numerous (in species and numbers) marine crustaceans. Millions of these minuscule (about 1/16th inch and less) creatures float in Florida's saltwater.

Arrow worms (*Sagitta* spp.) are transparent, torpedo-shaped little plankton that prey on copepods, larval fish and zooplankters. They spend their entire lives drifting. Spherical eyes allow them to see in every direction, right through their clear heads.

With a quick flip of their tails, arrow worms dart like arrows toward prey. With their powerful jaws, they grab victims as big as they are, and swallow them whole. The arrow worms, in turn, become a two-course meal for jellyfish, comb jellies and larger fish. And so it goes.

"Plankton, larvae, and fish may be transported from their home waters to regions where they would not ordinarily survive, temporarily buffered from the alien waters by the whirling remnant of the Gulf Stream. It has even been suggested that if the mini-ecosystem of a ring changes slowly enough and persists long enough to allow several generations of enclosed species to develop, the species may adapt, through natural selection, to the environmental conditions of the new region."
- H. S. Parker in *Exploring Oceans*

The Nature of Florida's Ocean Life

Free Spirits

Ocean sunfish (*Mola mola*) are also called head fish because they look like someone cut them in half, leaving only a huge floating head behind. They can grow to 13 feet and weigh 4400 pounds. Their skin is more than two inches thick and covered in slime.

Anglers fishing around the Gulf Stream see sunfish "sunning" on the surface of the open water. The sunfish lazily drift undisturbed in the warm water, secure that anglers aren't interested in their worthless meat. They are strong swimmers when necessary and, even though they're considered surface fish, can swim down to at least 1000 feet.

A female sunfish carries at least *300 million* eggs. When the tiny larva emerges, it's about the size of a raisin and looks more like a cross between a pufferfish and a sailfish, as shown to the right. The fish designer must have had a real giggle with this one.

Sunfish feed on jellyfish, **ctenophores** (pronounced "ten-oh-fours") and **salps**. Ctenophores are transparent, free-floating marine animals often mistaken for jellyfish. Sea walnuts and comb jellies are ctenophores. Shown below is a ctenophore called a **Venus' girdle** (*Cestum veneris*).

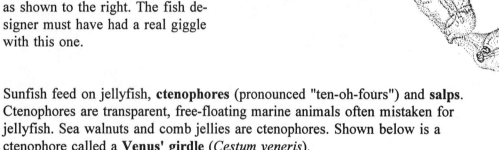

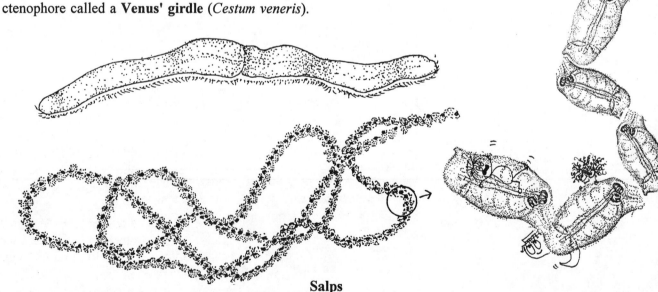

Salps

The **salps** shown above (*Thalia democratica*) remind me of jellyfish, but they are actually related to tunicates such as sea squirts. They are clear, barrel-shaped complicated creatures, only about an inch long, but with a highly complex reproductive system. Some of the youngsters develop within the mother, while others are tadpole-like larvae that form new colonies immediately. Adult salps connect with each other in long undulating chains of clear bodies, visible only by their guts as shown above.

"To partially stave off predation, the first, last, and only definite chance an animal has to hide in the open ocean is in water itself, and to hide in *water*, an animal must look like water. To look like water, it must be transparent. In mass, salps resemble nothing more than jellied water ... the heart, stomach, and intestine, contained in a package resembling a bung, are located on the lower edge of the barrel."
- John and Mildred Teal in *The Sargasso Sea*

Jellies

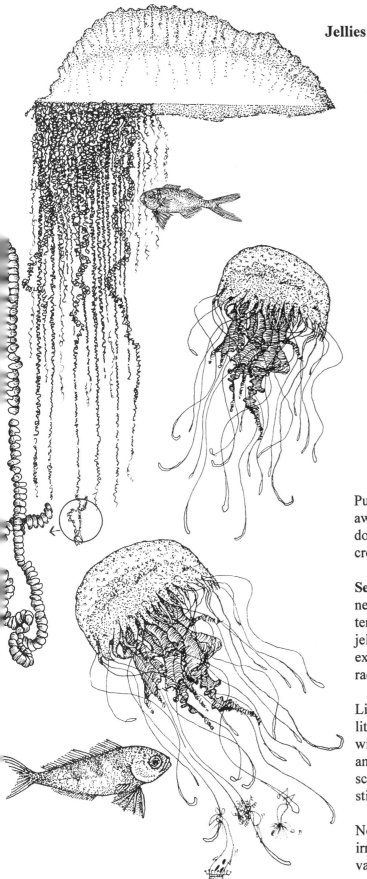

The floats of **portuguese-men-o-war** (*Physalia physalia*) look like the helmets of early Portuguese conquistadors, *men of war*, hence the name.

These gas-filled jellies float on the ocean's surface drifting with the Gulf Stream and winds. Why don't they dry out in the hot sun? They tip from side to side several times a day to keep their floats moist.

Their tentacles contain powerful stinging cells. When these microscopic cells are touched, say, by a swimmer's leg, a trigger goes off as quick as a jack-in-the-box, opening a trap door and releasing a long, coiled tube, as shown below. The tube is barbed and has toxic chemicals inside. When the tube touches our skin, it releases the strong toxin which causes pain, redness, and swelling.

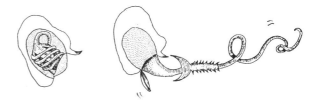

Putting meat tenderizer on jellyfish stings takes the pain away. Why? Using meat tenderizer on a welt breaks down protein bonds, therefore disrupting toxins that create the sting.

Sea nettles (*Chrysaora quinquecirrha*) (shown left) float near the surface of water in huge numbers, with 40 tentacles extending below each bell-shaped body. These jellies can be almost a foot wide with feeding tubes extending four feet below their body. Thin red spokes radiate from their clear, umbrella-like bodies.

Like Portuguese-men-o-war, sea nettles co-exist with a little fish, (*Nomeus gronovii*) that swims unharmed within the tentacles. However, Jack Rudloe, biologist and author of *The Erotic Ocean*, notes that when he scoops them in a bucket, the fish dies quickly from the stinging slime of its disturbed host.

Normally, a sea nettle's sting will only cause a mild irritation to humans, but the potency of their stings varies with the season. I wonder why.

"Despite doctor's orders, I'm not wearing sunblock. I've come across some information recently that suggests that petroleum-based sunscreens react photo-chemically with strong sunlight, releasing toxins which may trigger the very cancers they are supposed to prevent .. it is not a good feeling to fear the sun."   - William Thomas in "Sea Change" in *Ocean Realm* Magazine

The Drifters

Whales, dolphins, sharks and sailfish are the glamorous superstars of the sea, but for me, the smaller creative characters are the ones deserving our attention. Exotic relationships develop when marine creatures are brought together by the whim of a current or wind. Snails, slugs and jellies drift on the surface, bumping, touching, nibbling, protecting and sometimes eating each other.

Some of these smaller marine creatures literally go with the flow of the Gulf Stream. They use an easier and softer way to move through rough currents, letting the power of the Stream take them where they need to go.

Purple sea snails blow bubbles from mucus, creating a raft of bubbles to keep them afloat.

"Evolution ran riot in the ocean. The incredible end-products of nature's ingenuity and the bizarre forms they took are innumerable, but the purple sea snail must be one of the most prized exhibits ... it periodically ejects masses of actively swimming larvae, with shells less than 1/300 of an inch in diameter, able to feed immediately on the plankton of the surface waters until they are large enough to build their own rafts and prey upon jellyfish ... consider the *Janthina*'s method of transportation. Other sea snails cling orthodoxly to floating objects and therefore have no control over their movements. But *Janthina* fashions a raft of bubbles, from which it hangs upside down."
- Richard Perry in *The Unknown Ocean*

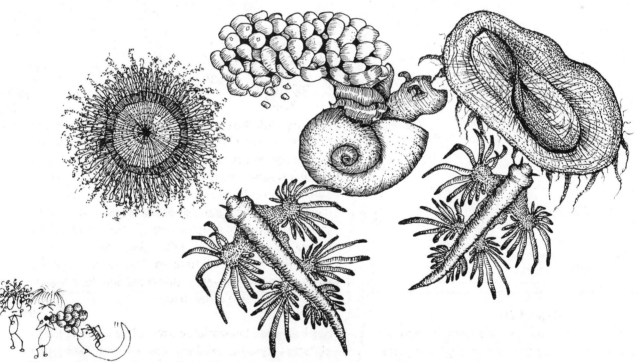

Shown clockwise from left are **blue buttons** (*Porpita porpita*), **purple sea snail** (*Janthina janthina*) and **by-the-wind-sailor** (*Velella velella*), with **glaucus** (*Glaucus atlanticus*) in the center.

A glaucus is a mollusk with no shell. The inch-long glaucus floats upside down clinging to the *underside* the surface of the water. They eat the tentacles of toxic creatures such as blue buttons, absorbing their poisonous cells to use for their own defense. Blue buttons are commonly found on our beaches, but by the time they reach our shore, they've lost most of their long, fragile tentacles to predators or breakage. The glaucus and the violet sea snail munch on the underside of jellyfish called by-the-wind-sailors.

"'Stream' is too humble a term for the mighty river in the sea ... the Gulf Stream is the most glamorous and best-studied component of the North Atlantic circulation. It's fully deserving of this attention."
- Henry S. Parker in *Exploring the Oceans*

Gulf Stream Passengers

Porcupine fish (*Diodon hystrix*) swallow air and water to puff up; not only does this make them seem larger to predators, it forces their spines to stand erect, deterring predators. They use their strong beaks to crush mollusks, urchins and crabs.

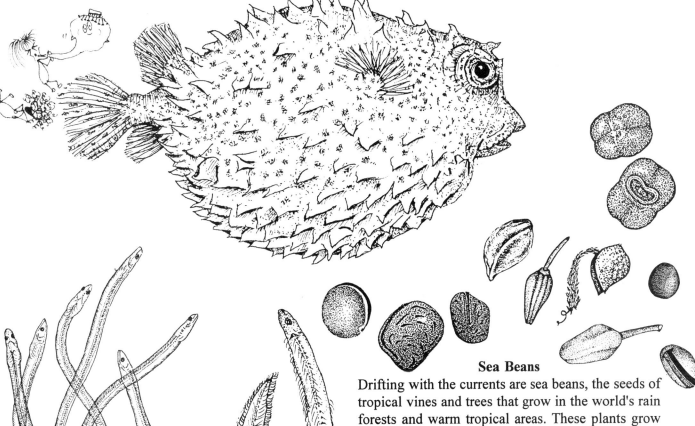

Glass Eels
Some of the creatures that use the Gulf Stream for transportation are eels as they go through the metamorphosis stage of changing from leaf-like larvae to immature eels. They are called *elvers* during this stage, but are also known as "glass eels" because their backbones and internal organs are clearly visible through their transparent skin. As the Gulf Stream swings by the edge of the Sargasso Sea, juveniles, such as these eels, get swooped up for the ride of their lives.

Sea Beans
Drifting with the currents are sea beans, the seeds of tropical vines and trees that grow in the world's rain forests and warm tropical areas. These plants grow along estuaries such as the Orinoco and Amazon Rivers in South America. When the seeds fall from their parent plants into the estuaries, they drift until they take root, rot or get eaten. Or they might flow through an inlet which takes them to the ocean. Those that survive continue to drift in saltwater for weeks, months or years, protected by their waterproof and durable outer coat.

Sea beans use the world's currents like a public transportation system, catching one current to another until the powerful Gulf Stream picks them up and carries them to a shore far from their origin.

Sea beans know nothing about social, political, cultural or geographic boundaries. They use the Gulf Stream to carry them to Cuba as readily as Palm Beach or Nantucket. Or they bypass North America entirely to drift slowly to European coasts.

"Tropical seeds and fruits (disseminules or sea-beans) which drift to temperate beaches and become stranded ... are messengers from exotic lands; for men of the sea, they represent victory over an ancient foe; for the superstitious, they represent gifts from the gods; and for the botanists, they are the end product of a plant dispersal mechanism."
- Dr. C. R. Gunn and John V. Dennis, Sr. in *World Guide to Tropical Drift Seeds and Fruits*

Fastest Fish in the Ocean

The outer edge of the Gulf Stream is often clearly marked by an assortment of debris caught at the intersection of the fast moving waters and the stationary ocean to the west. Large and small fish are attracted to this accumulation, lured by the potential for food.

Sailfish Alley and the Fastest Fish in Our Ocean

Sailfish (*Istrophorus platypterus*) show up in December along Florida's "Sailfish Alley," an 80-mile road in the ocean between Sebastian Inlet and Palm Beach where sailfish consistently swim in large numbers. This road steadily widens the farther south you go. Here, the Gulf Stream flows one to five knots, and a 120-foot deep contour follows the edge of the Continental Shelf. Sailfish congregate at this intersection of clear blue Gulf Stream water and green coastal ocean, attracted to the baitfish which gather around the edge.

Rock communities and the irregular bottom depths of Sailfish Alley provide hiding places, food and breeding grounds for the smaller fish that sails feed on. Small tuna, jacks, squid, ballyhoo, goggle-eyes, pinfish, puffers, octopodes, triggerfish, mullet and flying fish all use this area like a flea market, all shopping for something different. Some get sucked in themselves, while others swim away with a grand prize.

Charging the Ball

As a team, a group of sails will work together to herd schools of small fish into a compact ball. The sails continue to circle this bait-fish ball until it's tight and dense. The sails then charge the ball, slashing through it with their bills; the small fish -- stunned, maimed or killed -- are eaten by the marauding sails.

Larval stages of the sailfish, shown below (greatly magnified), don't look much like the adult billfish they become.

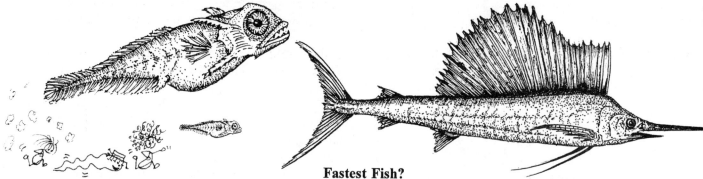

Fastest Fish?

Ichthyologists have clocked sailfish swimming 68 miles per hour, declaring them the fastest fish in the world.

But Not So Fast

"Not so," says Harpo Katz, professional angler, who disputes the claim that sailfish are the fastest fish in the world. "I've burned many a thumb from the *real* fastest fish in the ocean -- as we all know, it's the wahoo ... hence their name: wahooooooooo!" (*Acanthocybium solanderi*)

The International Fish Association agrees with Angler Harpo: According to *World Record Game Fishes*: "Wahoo are reputed to be one of the fastest fish in the sea ... the first scorching run of a hooked wahoo may peel off several hundred yards of line in seconds, and the heat generated by the friction has been known to burn the drag on some reels."

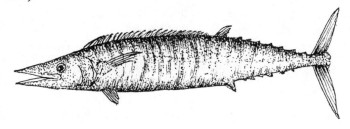

"The more one thinks about the Gulf Stream the more it becomes apparent that this is a feature of our earth with a two-fold meaning ... the observable effect of a fast-moving body of water transversing the North Atlantic ... and the more nebulous fact that the Gulf Stream is only part of a whole, a general circulation of water in the ocean, which is inexorably determined by the rotation of the earth together with its atmosphere."   - Thomas F. Gaskell in *The Gulf Stream*

Disguises in the Gulf Steam

The Gulf Stream is the Atlantic's answer to a subway system, picking up passengers and dropping them off miles from their origin. One of its regular passengers is the squid.

Squid are the most sophisticated of all the invertebrates, with an intelligence that surprises anyone watching their ability to hide, attack and deceive. They can change color and texture to match their surroundings in seconds. And, to avoid predators, they can jet backwards by shooting water through a siphon in their mantle (the fleshy fold that covers their organs). These jet setters are known as the fastest creatures in the ocean for short distances.

Ink Spots and Smoke Screens

A squid can also squirt ink -- not only to distract predators long enough to make an escape, but also to mask their odor. The ink doesn't spread through the water, but stays roughly the same size and shape as the squid. Simultaneously, the squid becomes pale, so a predator is more likely to go after a noticeable dark shape than the ghost-like hint of a squid.

Sometimes the squid will squirt a more watery ink which *does* spread through the water -- this is used as a smoke screen to hide the squid as it makes its escape.

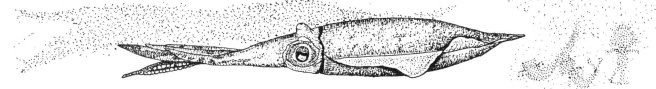

Another marine creature able to camouflage itself is the **gulf Stream flounder** (*Citharichthys arctifrons*). Flounders lie flat on the ocean's bottom, their color and texture camouflaging them in the surrounding sand.

These flounders are relatively drab, small fish (less than seven inches), but an interesting characteristic makes them worth mentioning. Within days of their birth, the eye on one side of their head migrates until it's next to its twin on the other side. This is a typical trait in most flat fish.

Flounders live on the sandy bottom and only spend time on the surface as larvae. Eggs float to the surface and become larvae, living on plankton. They slowly move back to the bottom as they mature, changing their diet to worms and crustaceans. Their young bodies are fish-like but gradually go through dramatic changes: in addition to the eye migration, they change shape to adjust to life on the ocean floor, orienting themselves horizontally.

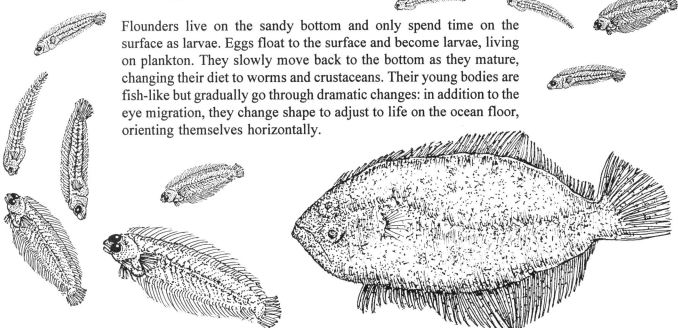

"Exotic tropical fish, mollusks, starfish, and many others have settled here hundreds of kilometers north of their usual haunts. Most of them arrived from the Caribbean via the Gulf Stream and fortuitously found this last jumping off place on the shelf before the warm current veers easterly into the deep North Atlantic."

- John L. Culliney in *The Forests of the Sea*

The Nature of Florida's Ocean Life

SARGASSO SEA

"But the Sargasso is not nearly as daunting or menacing as its legends, which describe it as an 'island of lost ships.' In reality, it is a calm, lazy region of the North Atlantic Ocean, the 'eye' of the ocean's clockwise circulation pattern."
- Steven Ginsberg

Sargasso Community

Inside the powerful circle of Atlantic currents is a *huge* expanse of warm water, called the Sargasso Sea. It's an enormous, slow-moving area of calm sea -- an international meeting place, bordered by the Gulf Stream on the west and the Azores to the east. All boundaries are undefined, unpredictable and irregular.

Its surface is a thick layer of seaweed: a marketplace of two million square miles supporting an exotic culture of crabs, shrimp, fish, snails, beans, turtles and human-made debris.

Below the weed, a mysterious underground network of deep sea creatures live, conducting rituals of survival with a secretive self-assurance. Most creatures, including humans, don't have the sophistication needed to barter in this relatively unexplored habitat.

The Weed

The mass of seaweed that floats on the water's surface is undisturbed by the surrounding currents. **Sargassum weeds** (*Sargassum natans and S. fluitans*) differ from other marine plants in that they are free-floating. Rough estimates indicate *seven million tons* are floating in the Sargasso Sea, drifting loosely clockwise from the east side of the Gulf Stream to the middle of the Atlantic.

Sargassum comes from the Portuguese word *sargaço* meaning grape. The small air bladders holding the weed afloat look like little yellow grapes. Although several types of sargassum exist, two of them are commonly found off Florida's coast. The difference between the two is mostly in the little air bladders: the bladders of *Sargassum natans* have tiny spears projecting from them, which the fuller bladders of *Sargassum fluitans* don't have.

As it floats, swirls, and separates, patches of weed are picked up by the Gulf Stream, adding to the hodge-podge of ocean junk in this conveyor-belt current. The Gulf Stream carries the weed with its accumulated debris to our shore, where the high tide creates a *wrack*, a line of beach debris composed mostly of this seaweed.

When I see sargassum wash onto shore during my beach walks, I know this is the end of its life, but I also know that it brings food from the sea for lots of our shore birds, crabs, insects and raccoons.

"Here on the deep watery plain, one readily perceives the frailty of humanity but also learns to purge the small unreasoning fears that come with too much glancing over the shoulder at distant shapes."
- John Culliney in *Forests of the Sea*

Sargasso: The Ultimate Cruise

Patches of sargassum weed travel like cruise ships carrying parties of fish, crabs, slugs, and sea turtle hatchlings. Among the passengers are inanimate objects that every cruise ship needs: jetsam and flotsam of imported food, bottles, toys, sneakers, ping pong balls and light bulbs.

Jetsam and Flotsam

What's the difference between jetsam and flotsam? Cargo or equipment *thrown overboard* (jettisoned) to lighten a ship's load is usually called jetsam. Typically, jetsam sinks or is grounded. Flotsam, however, is an "accumulation of unimportant, disordered, and miscellaneous trifles." But the "trifles" that gather in the seaweed are not such trifling matters as I once thought.

Garbage, tar, toys, electronics, bottles, furniture and balloons all become part of this mobile habitat. Some of this trash provides homes and shelters for barnacles, fish and crustaceans.

> "All the flotsam and jetsam of the ocean -- drifting tree trunks, boxes, light-bulbs -- quickly become the encrusted homes of small oceanic animals. But the most extraordinary of these microscopic island havens on the shelterless plains of the ocean are blobs of fuel oil, which become populated not only with barnacles and hydroids and egg-laying *Halobates* but, most remarkably, with a small fish which almost exactly resembles the oil blob in both shape and color. Since fuel oil has been floating on the oceans of the world for only about forty-five years, this fish cannot have evolved this resemblance in so short a period."
> - Richard Perry in *The Unknown Ocean*

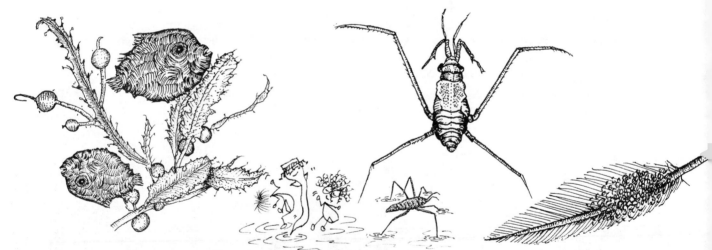

> "The **water strider**, *Halobates*, is a skin-sucking sea skimmer. Their name literally means 'salty walkers' from the Latin *hal* for salty and *bates* for walker. The only true saltwater-inhabiting insects, or, as Jim Ingraham says, 'The only bugs worth their salt.'"
> -Dr. Curtis Ebbesmeyer in *All Things Afloat*

The halobate's tiny feet make a dent in the ocean's surface as it rows with its middle pair of oars. Dr. Ebbesmeyer explains that the ocean's 'skin' is made of five layers, all totalling a thickness of one millimeter:

> "Halobates not only walks upon the sea's thinnest, uppermost skin layer, but also feeds from the strata beneath ... and they literally talk with one another by dancing rhythms on the sea's trampoline -- a sort of Morse code beat on a drum tightly covered with sea skin."

The water strider can lay up to 20,000 eggs on a single floating feather.

"Why do we all work so hard? It's because we think we need so much stuff." - Steve McLeod

Rain Forests in the Sea?

What does the Sargasso Sea, in the middle of the Atlantic Ocean, have in common with the world's rain forests?

Sea beans, growing in rain forests around the globe, drift in ocean currents. Some of them reach the Sargasso Sea, collecting barnacles, lacy crusts and minute algae along the way. They can travel like this because of dispersal mechanisms other seeds lack: a hard outer shell that protects them from saltwater and an air cavity inside that provides buoyancy.

What's the Point?

Nature has provided incredible gifts of neatly packaged seeds that can travel with the world's currents to other continents -- but why would a seed travel to a continent whose soil and climate aren't favorable for growth?

The answers aren't all in yet, but suppose the chemicals in these seeds provided substances lacking on non-native soils? Or suppose these seeds could provide, as suspected, cures for Alzheimer's, diabetes, cancer, AIDS and other diseases? What if the seeds could start new forests in a future Earth with different climates and continents?

Paul Alan Cox, Botany professor and Dean at Brigham Young University sees the rain forests and its healers as holders of cures for many of the world's diseases. He suggests, "We Westerners have to suspend judgement at these times. Look at our own belief in doctors wearing white coats."

Dr. Cox states, "I see ethnobotany -- the study of the relationship between people and plants -- as the key to the preservation of this vast collection of species as well as a pathway to halting many diseases ... Indigenous people have been testing plants on people for thousands of years."

Sharing Experience and Hope

But figuring out which rain forest plants to study is difficult and time-consuming among the thousands of candidates. Twelve years of research and up to $300 million are needed, traditionally, to study each new drug. Rain forest healers who have passed on plant knowledge from generation to generation are now being consulted for information and advice. Using the wisdom and experience of the rain forest residents is proving to be more effective than random sampling.

Willing to go to Any Lengths

The **hamburger bean** (*Mucuna* spp.) grows inside a prickly pod dangling on a long stem. The pod's sharp "hairs" are toxic -- a deterrent against rain forest predators such as small rodents. Even if an animal manages to climb the long stem, it still must get past the prickly pod for the bean inside. And then, even if it gets through the poisonous outer package, the animal still has to contend with the rock-hard coating of the seed. That's a lot of work for a bean. What's inside to make it so desirable?

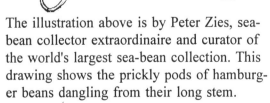

The illustration above is by Peter Zies, sea-bean collector extraordinaire and curator of the world's largest sea-bean collection. This drawing shows the prickly pods of hamburger beans dangling from their long stem.

"Sea beans have been used by ancient navigators as indicators of lands beyond the horizon, as God's gifts to be used in times of great need, and as objects of wonder or as ornaments."   - C. R. Gunn and J. V. Dennis, Sr., in *World Guide to Tropical Drift Seeds and Fruits*

Sea Bean Soup

For centuries, people reasoned that if sea beans came from the sea, they must, therefore, grow in the sea. Without the knowledge of currents and distant land, why would anyone think differently? Some of the sea beans these early explorers found are described below.

To find a sea bean in an ocean swirling with white caps, sea weed, tar blobs, turtles, and trash is almost impossible, but sailors in past centuries have recorded seeing sea hearts and some of the larger drifters: calabashes, coconuts and box fruits.

Sea hearts (*Entada gigas*) are thick, brown seeds that grow in long pods in Central and South America. The pods, sometimes a yard long, each contain a dozen or so of the hard dark seeds. When the pods dry out, the seeds fall into the river system and flow out to the oceans. Currents pick them up and carry them to Florida -- or they may eventually drift to European shores.

Floridians often find **coconuts** (*Cocos nucifera*) washed up on the beach. Florida is not the only shore that receives these drift seeds from tropical countries. Western coasts of Europe, especially the United Kingdom, regularly report drift seed activity, with coconuts drifting to Ireland and Scotland.

Spanish and Portuguese explorers thought the coconut, with its three small holes, looked like a human face, so they called it *coco* which means grinning or grimacing face.

Calabashes (*Crescentia cujete*), one of the larger drifters, range in size from a small orange to a large grapefruit. Calabashes grow on 30-foot evergreen trees in most of the tropical American countries.

Tiny seeds inside are packed tightly together. When this seed-wad dries, the calabash rattles when shaken ... hence, calabashes are used to make maracas. Because of their hard and durable "shell," calabashes are also used as bowls, bailers or ladles.

Box Fruits (*Barringtonia asiatica*) are fibrous, lantern-like drifters that have been used as fishing floats and decorations. While still on the tree, the fruit is a smooth box, round on top, with the bottom ending in a pony-tail of pale strands. By the time it reaches a shore far from its country of origin, it has lost most of its hard outer casing and looks like a ball of tangled beige twine.

Anglers on some islands pulverize box fruit seeds and use the powder to catch fish by sprinkling it in their fishing areas. The poisonous powder contaminates the water and stuns the fish which are then easily caught. Humans are unaffected by the chemical.

"It must not be thought that scientific beachcombing is anything new. Columbus is said to have strengthened in his view that land lay to the west by hearing about alien objects stranded on beaches in the Azores. The Columbus bean, *favas de Colom*, which we know as the sea heart, is said to have been one of them."   - John V. Dennis, Sr., in *Beachcombers I Have Known and Their Hobbies*

Fish in the Sargassum

Below the thick layer of sargassum, a dream-like cast of creatures exist whose survival skills display more creativity and adventure than any Spielberg movie could ever script.

Sargassumfish (*Histrio histrio*) is a master of camouflage. It stares at the world hidden behind a protective jungle of leaves, lumps and colors that match his own. Even when a predator approaches dangerously close, the chances of it seeing anything more than a salad, are unlikely.

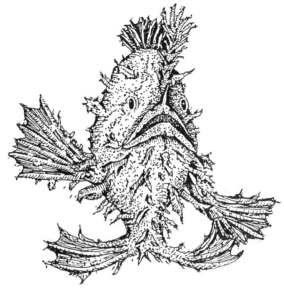

The sargassumfish crawls around the weed on its fins, so carefully that a victim doesn't realize it's being stalked. When the sargassumfish's mouth is right next to its prey, it takes action: its jaws open so wide and fast that a powerful suction vacuums the prey straight down its throat.

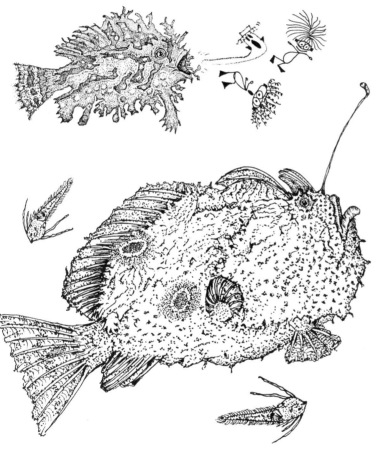

One of the sargassumfish's cousins, the **angler fish** (*Antennarius ocellatus*) is also known as a frog fish. It uses its fins to crawl and hop over the surface of the seaweed.

This beautiful creature has devised a way to stay hidden even while hunting for a meal. Victims are lured toward the frogfish's mouth by a fake worm wiggling from its built-in fishing pole. The "worm" is actually a fleshy blob that lures passersby to the angler's big mouth which sucks in food swimming too close.

The "fishing pole" is a modified first spine of their dorsal fin. When the pole isn't in use, the angler fish tucks it between its eyes. Female anglers have an added attraction: their bait glows. Who could have made this fish any more wonderful?

The stripes and coloring allow the little **sargassum pipefish** (*Syngnathus pelagicus*) to disappear among the weeds and the other sargassum characters.

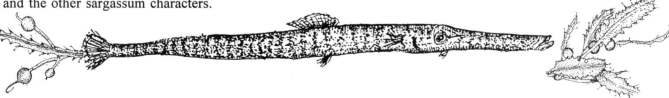

"Nature serves as a good example of how trial and error can be used to make changes. Every now and then genetic mutations occur -- errors in gene reproduction ... occasionally a mutation provides the species with something beneficial and that change will be passed on to future generations. If there had never been any mutations from the first amoeba, where would we be now?"
- Roger von Oech in *A Whack on the Side of the Head*

More Fish in the Sargassum

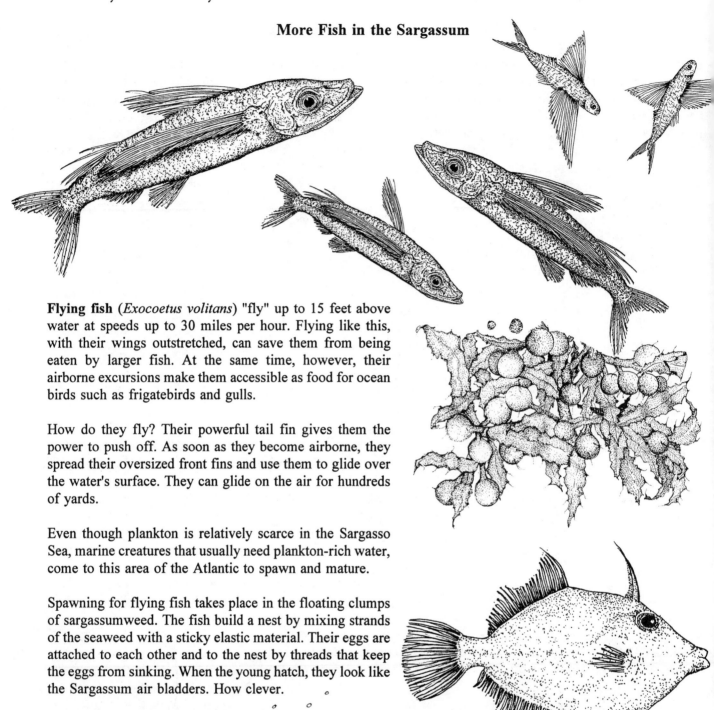

Flying fish (*Exocoetus volitans*) "fly" up to 15 feet above water at speeds up to 30 miles per hour. Flying like this, with their wings outstretched, can save them from being eaten by larger fish. At the same time, however, their airborne excursions make them accessible as food for ocean birds such as frigatebirds and gulls.

How do they fly? Their powerful tail fin gives them the power to push off. As soon as they become airborne, they spread their oversized front fins and use them to glide over the water's surface. They can glide on the air for hundreds of yards.

Even though plankton is relatively scarce in the Sargasso Sea, marine creatures that usually need plankton-rich water, come to this area of the Atlantic to spawn and mature.

Spawning for flying fish takes place in the floating clumps of sargassumweed. The fish build a nest by mixing strands of the seaweed with a sticky elastic material. Their eggs are attached to each other and to the nest by threads that keep the eggs from sinking. When the young hatch, they look like the Sargassum air bladders. How clever.

Orange filefish (*Aluterus schoepfi*) spend their juvenile years within the Sargasso community. They can grow to two feet long, but typically are much smaller. Their unusual pelvic bone, with its graceful curve, was used by fishermen as letter openers.

Do fish pass gas? Yes. Fish develop gas in their guts and get rid of it by excreting a thin gelatinous tube with the gas and their poop all neatly disposed of in one package. Sand tiger sharks (*Eugomphodus taurus*) not only pass gas, they *use* it as a buoyancy device. "The shark swims to the surface and gulps air, swallowing it into its stomach. It can then fart out the required amount of air to maintain its position at a certain depth."

-Alexandra Osman

Designer Creatures in the Sargassum

Sargassum crabs (*Portunus sayi*) are usually less than two inches. Their brownish-yellow bodies are spotted with white to give them the "sargasso look."

During its evolution, the sargassum nudibranch or **sea slug** (*Scyllaea pelagica*) lost the hard shell of its snail ancestor. Shell-less, it is now more agile and can glide easily through water. The creamy-orange body has brown and white patches and flecks of rust to let it blend with the surrounding weed. These beautiful creatures are usually less than an inch and are sometimes found tangled in the clumps of seaweed that wash onto Florida's beach in the fall.

♪ ♭ ♪ ♮ ♪ ♫ ♪ I Ain't Got No Body ♫ ♪ ♯ ♪ ♮ ♫

Sargassum sea spiders (*Endeis spinosa*) are related to horseshoe crabs and land spiders. On the mats of sea weed, they walk from strand to strand and from leaf to leaf on their long legs. The body of our more familiar land spiders have two parts, while this sea spider barely has a place for the legs to connect -- it has no body.

So where are their organs? With no body to hold their internal organs, they're located in the long, thin legs. What about a pregnant sea spider? Where does she put her eggs? Females lay eggs in masses; her partner collects the eggs and carries them on his special egg legs called *ovigers*. From this mass of eggs, larvae emerge which stick with Dad until mature enough to separate and start their own bodiless lives. Fully mature, they are usually no larger than a lady bug.

Sargassum shrimp (*Latreutes fucorum*) are only about ½-inch and colored to match the sargassum weed.

"One of my favorites are the sargassum shrimp. One day, one of the shrimp was trying to eat a coquina and got his paw stuck between the coquina's valves. The poor shrimp limped all over, shaking his claw to try to get the coquina to let go. He finally succeeded, thank goodness."   - Cathy Yow

Loggerhead Sea Turtles

Enormous female loggerheads come to Florida's beaches each summer to lay their eggs. About two months later, the hatchlings emerge from the sand to begin their life at sea. Where do they go when they leave Florida?

Each summer, I watch hundreds of the two-inch hatchlings scurry from their clutch toward the ocean -- I flinch whenever I see them tumbling wildly in the foamy surf -- but then they disappear into the dark ocean water. Most of them don't survive the obstacle course of raccoons, birds, crabs, sharks, barracudas, rough tides, heat, bacteria. But the ones who do survive -- wow -- what a ride as they travel with the Gulf Stream to the Sargasso Sea.

They travel about 25 miles nonstop in 24 hours, swept along for part of their journey by currents. When they reach the first clumps of sargassum weed, I imagine them climbing aboard, finally settling onto their floating homes which will feed and entertain them for many years.

The weed mats drift northeast, clockwise with the flow of the North Atlantic Gyre, giving the hatchlings a new view each day and plenty of fresh food. The sargassum also provides the little sea turtles with a smorgasbord to graze on -- tiny fish, eggs, barnacles, jellyfish, shrimp and all kinds of larvae. The weed itself is a salad bar of nutrients for them.

For the next 10 to 12 years, the hatchlings drift with the weed as it circulates thousands of miles, carrying its hitch hikers to the eastern edge of the Sargasso Sea. Tagging and DNA testing have confirmed that turtles in the Azores, thousands of miles east of the United States, were actually born in Florida.

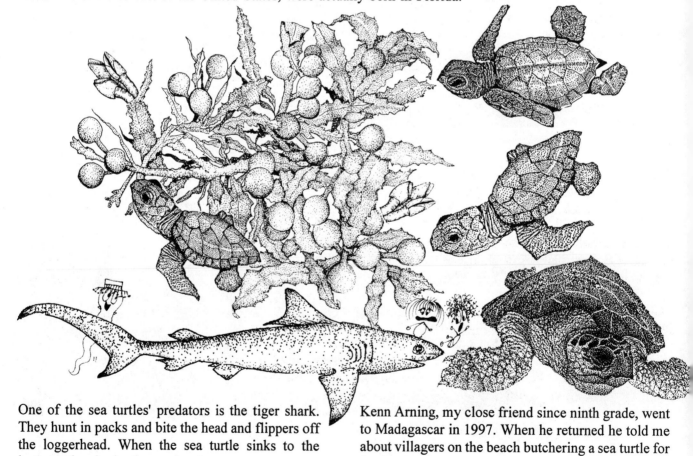

One of the sea turtles' predators is the tiger shark. They hunt in packs and bite the head and flippers off the loggerhead. When the sea turtle sinks to the bottom, the sharks devour her meat.

Kenn Arning, my close friend since ninth grade, went to Madagascar in 1997. When he returned he told me about villagers on the beach butchering a sea turtle for food. His story ended with a moment of silence, and then he said, "It fed the villagers for a week."

"Our nature lies in movement; complete calm is death."
- Blaise Pascal in Pensées

For Whom the Light Glows

The average depth of the Sargasso Sea is about three miles. This deep area in the Sargasso Sea holds an array of marine creatures that travel vertically as easily as horizontally.

Squids descended from mollusks whose outer shell was necessary for protection and survival. For some squids, the shell became a hindrance, and most of them lost their protective outer covering through evolution. One little **squid** (*Spirula spirula*), however, evolved so that its shell is now *inside* its fleshy body.

This deep sea squid, less than two inches long, lives below the sargassum weed. Like other squids, they have eight arms and two tentacles. The other end of the animal has two small fins with a circular disk in between. A bead-like organ in the middle of this disk emits a steady, yellowish-green light which glows for hours. Their normal position is vertical with their arms hanging down, so the disk is like a beacon on top -- for whom does their light glow?

When the squid dies, their internal skeleton rises to the surface and drifts with the seaweed, beans, trash, and other sargassum creatures. These skeletons, known as ram's horns or spirula, drift in the ocean until currents and tides carry them onto a beach thousands of miles from their origin. Beachcombers on five continents report finding these spirula "shells."

The shell is divided into 25 to 37 gas-filled chambers. How can this fragile shell remain intact in water where the pressure is nearly half a ton ... *and* travel almost a mile to the surface without breaking? As the squid rises, gas is slowly discharged from the chambers, keeping it in equilibrium with the water pressure surrounding it.

Calamari, the seafood eaten with linguine and other dishes, is made from **opalescent squid** (*Loligo opalescens*), a little squid that grows to about seven inches.

This squid shoots through the water, aiming at prey with torpedo-like precision. It grabs its prey with special suckers at the end of its two long tentacles.

Females lay transparent eggs from which tiny replicas of the parents are born, each smaller than a pencil eraser.

"Now that I know at least one wonderful gift comes each time I encounter chaos, I encourage rather than avoid chaos. My days have become as fluid as a squid in the Sargasso Sea."   - Anonymous

The Nature of Florida's Ocean Life

Deep Sea Changes

In addition to eels, many other deep sea creatures develop with the sargassum weed and change drastically in appearance during their lifetime.

One of these marine creatures that changes physically is the **pelagic clam worm** (*Nereis pelagica*). These iridescent-green-brown worms develop huge eyes as they begin breeding in the deeper, open water.

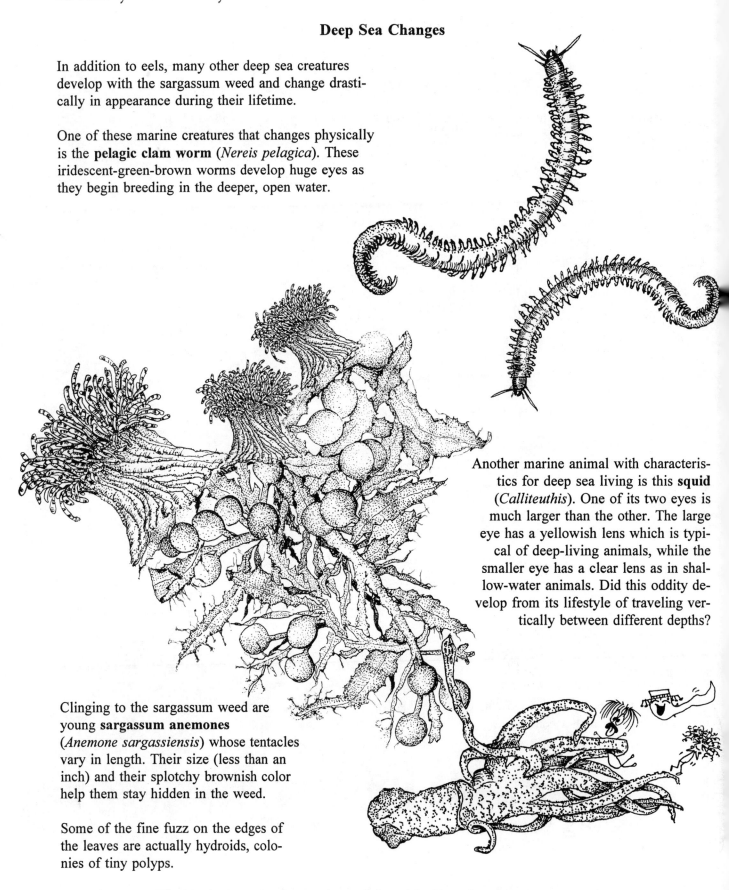

Another marine animal with characteristics for deep sea living is this **squid** (*Calliteuthis*). One of its two eyes is much larger than the other. The large eye has a yellowish lens which is typical of deep-living animals, while the smaller eye has a clear lens as in shallow-water animals. Did this oddity develop from its lifestyle of traveling vertically between different depths?

Clinging to the sargassum weed are young **sargassum anemones** (*Anemone sargassiensis*) whose tentacles vary in length. Their size (less than an inch) and their splotchy brownish color help them stay hidden in the weed.

Some of the fine fuzz on the edges of the leaves are actually hydroids, colonies of tiny polyps.

"The scientist does not study nature because it is useful ... He studies it because he delights in it, and he delights in it because it is beautiful."
- Jules Henri Poincaré, French scientist

Eels in the Sargasso Sea

"The whole story of the eel has come to read almost like a romance,
wherein reality has far exceeded the dreams of fantasy."
- Johannes Schmidt (written in 1912 in the scientific journal "Nature")

Millions of mature freshwater eels from rivers in America and Europe migrate to the Sargasso Sea to spawn each year. Their migration begins during the new moon in autumn when the nights are dark enough to hide them from predators. Their drive to migrate is so strong that they will even cross over dried up river beds to get to the big spawning convention in the Sargasso Sea.

American and European eels deposit eggs in the Sargasso Sea more than 1000 feet below the thick seaweed mats, and then die.

After hatching, the clear, leaf-shaped larvae (called *leptocephalus* meaning 'skinny-head') drift with the Gulf Stream toward their ancestral coasts.

"Their young, born in the deep, with no parental guidance, but with astonishing, pre-programmed instincts, swim to the streams vacated by their parents and take up a fresh-water life until they, too, complete the cycle by returning to the Sargasso to spawn and die."
- Mildred and John Teal in *The Sargasso Sea*

During the drifting, the young eels begin to grow and gain some color. These smaller versions of their parents are called *elvers*, as mentioned in detail on pages 28 and 29.

American eel larvae reach our coast after about a year and are called elvers. By the time these elvers have reached Florida's coast, they are a few inches long. However, the European eels take three years to find their way to their parents' coast and, accordingly, their maturity is slower.

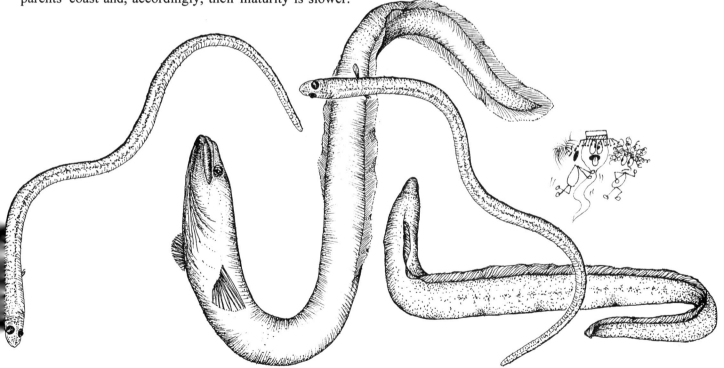

"Though millions of tiny eels hatch in spring and begin drifting toward Atlantic coastal rivers, the watery grave of the elders remains, at least for now, unmarked."
- Cheryl Lyn Dybas in "By the Dark of the Moon, Eels Slither out to Sea" in *The Washington Post*, October 9, 1996

El Niño

Weather conditions influence the Sargasso Sea and the Gulf Stream. We see the results when sea weeds and ocean junk from those areas are brought to our shore. Beachcombers love gusty winds that bring lots of drift material to the beach. Why then, don't we *always* have lots of sea beans, weeds and trash on the beach after high tides and strong winds? Why do these same forces occasionally sweep our beach aggravatingly clean?

Periodically, Florida doesn't get a fair share of drift material, as happened recently in 1983, 1994 and 1998. This can occur despite strong winds and tides that usually bring lots of beachcombing treasures. Of course, we blame El Niño for this and everything else, including too-tight shoes, bad eyesight, and not showing up for work on time.

So what is El Niño?

El Niño is an area of warm water in the Pacific. During normal weather conditions, southwest winds cause cold water to well up from the deep ocean. But in El Niño years, this doesn't happen. The result is an unusually warm current off the west coast of South America, causing floods, lack of fish, devastating winds and diseases. All species are affected, Florida's included ... as if an invisible force from the Pacific moves across the continent, pushing everything in its path. El Niño, which means *the little boy*, has a counter-force in the Atlantic called La Niña, *the little girl*. She's a trouble-maker too, but her energy subdues some of El Niño's intensity.

I have an image in my head of El Niño's powerful body -- a monster blob almost the size of America, getting fatter and meaner as it sucks up energy from the Pacific Ocean. As it grows larger, its mass works its way across the continent until it backs into the Atlantic. With its rear end hanging over Florida's coast, it keeps the debris in the Gulf Stream and the Sargasso Sea *away* from our coastline. If all the seaweed and debris are blocked from Florida by this big butt-head, *nothing* will be brought to our shore, no matter how high the tides are.

BUT! When this giant squatter finally moves, like a fat pig rising from a mud hole, an accumulation surges toward us, carrying all the treasures we didn't get during El Niño's tantrum. When this happens, we see the legendary *armada* of sea beans, accompanied by a fleet of toys, sneakers, weeds, messages in bottles ... and who knows what else? I expect the unexpected.

"All shall be well,
And all shall be well,
And all manner of things shall be well."
- Dame Julian of Norwich in the Thirteenth Century

"All is never said." -Togo saying
(Togo is a country in West Africa)

Bibliography

Audubon Society. *Field Guide to North American Fishes, Whales, and Dolphins.* N.Y.: Alfred A. Knopf, 1983.

Berrill, N. J. *The Life of the Ocean.* The World Book Encyclopedia/McGraw-Hill Book Company, 1966.

Burton, Maurice. *The New International Wildlife Encyclopedia.* Purnell Refernce Books, 1980.

Childers, Frank M. *History of the Cape Canaveral Lighthouse.* Melbourne: The Brevard Museum, 1983.

Culliney, John L. *The Forests of the Sea.* San Francisco: Sierra Club Books, 1976.

DeWire, Elinor. *Guide to Florida Lighthouses.* Englewood: Pineapple Press, Inc., 1987.

Gaskell, Thomas F. *The Gulf Stream.* New York: The John Day Company, 1972.

Gosner, Kenneth L. *A Field Guide to the Atlantic Seashore.* Boston: Houghton Mifflin Company, 1978.

Groves, Don. *The Oceans.* New York: John Wiley & Sons, 1989.

Gunn, C. R. and Dennis, J. V. *World Guide to Tropical Drift Seeds and Fruits.* New York: Quadrangle/The New York Times Book Co., 1976.

Headstrom, Richard. *All About Lobsters, Crabs, Shrimps and Their Relatives.* NY: Dover Publications, Inc., 1979.

Humann, Paul. *Reef Creature Identification.* Jacksonville: New World Publications, 1992.

Kale, Herbert W., and Maehr, David S. *Florida's Birds.* Sarasota: Pineapple Press, 1990.

Kaplan, Eugene, H. *Southeastern and Caribbean Seashores* (Peterson Field Guide). Boston: Houghton Mifflin, Co., 1988.

Littler, D. S. and Littler, M. M., Bucher, K. E., Norris, J. N., *Marine Plants of the Caribbean.* Washington, D.C.: Smithsonian Institution Press, 1989.

McCarthy, Kevin and Trotter, William. *Thirty Florida Shipwrecks.* Sarasota: Pineapple Press, 1992.

Meinkoth, Norman A. *The Audubon Society Field Guide to North American Seashore Creatures.* New York: Alfred A. Knopf, 1981.

Parker, Henry S. *Exploring the Oceans.* New Jersey: Prentice-Hall, Inc., 1985.

Robins/Ray/Douglass. *A Field Guide to Atlantic Coast Fishes* (Peterson Field Guide). New York: Houghton Mifflin, Co., 1986.

Rudloe, Jack. *The Sea Brings Forth.* New York: Alfred A. Knopf, 1968.

Rudloe, Jack. *The Erotic Ocean.* New York: Alfred A. Knopf, 1968.

Singer, Steven D. *Shipwrecks of Florida.* Sarasota: Pineapple Press, Inc., 1992.

Teal, John and Mildred. *The Sargasso Sea.* Boston: Little, Brown and Comapny, 1975.

Voss, Gilbert. *Seashore Life of Florida and the Caribbean.* Miami: Banyan Books, 1976.

Williams, Winston. *Florida's Fabulous Seashells and Seashore Life.* Tampa: World Publications, 1988.

Index

Algae, 23
Ambergris, 26, 27
Anemones
    Corkscrew, 45
    Onion, 17
    Sargassum, 68
    Sloppy guts, 17
    Tricolor, 45
Apollo Eleven, 50
Bioluminescence, 16
Birds
    Arctic terns, 24
    Common terns, 24
    Frigatebirds, 25
    Northern gannets, 25
    Gulls, 25
    Storm petrels, 24
Bottles, 11
Cameos, 36
Castro, Fidel, 10
Comb jellies, 17, 52
Copepods, 51
Coquina rock, 13
Coral, 34-35
    Staghorn, 35
    Star, 35
    Stinging, 35
    Soft (see Octocorals)
Crabs
    Hermits, 45
    Sargassum, 65
    "Stoned," 13
Crustaceans
    Crabs
        Hermit, 45
        Sargassum, 65
    Lobsters, 42-43
    Shrimp
        Cleaner, 44
        Ghost, 13
        Mantis, 44
        Pistol, 45
        Sagassum, 65
Ctenophores, 17, 52
Cuba, 10, 11
Dinoflagellates, 20
Drifting Seed Newsltter, 27
Duckies, 18
Echinoderms
    Feather stars, 40
    Sand dollars, 40
    Sea cucumbers, 41
    Star fish, 40
    Swimming crinoids, 40
    Urchins, 40
Eels (see Fish)
El Niño, 70

Fish
    Angler (Frogfish), 63
    Barracuda, 47
    Blennies, 36
    Eels, 28, 29, 55, 69
        Elvers, 29, 69
        Morays, 41
    Flounder, 57
    Flying, 64
    Great barracuda, 47
    Moray eel, 41
    Nomeus, 53
    Ocean sun, 52
    Oil blob, 60
    Orange file, 64
    Pearl, 41
    Porcupine, 55
    Pipefish, 63
    Queen parrot, 47
    Queen trigger, 47
    Sail, 55
    Sargassum, 63
    Shark, 66
    Sun, 52
    Trumpet, 47
    Wahoo, 56
Flotsam, 60
Franklin, Ben, 50
Fulgurites, 13
Glaucus, 54
Gulf Stream, 48-49, 70
Halobates, 60
Inlets, 20
"Insects"
    Halobates, 60
    Sand hoppers, 22
    Sea roaches, 22
    Sea spiders, 65
Janthina, 54
Jellyfish
    Blue buttons, 54
    By-the-wind-sailor, 54
    Comb jellies, 17, 52
    Portuguese-men-o-war, 53
    Sea nettles, 53
    Sea walnuts, 17
    Sea wasps, 51
    Upsidedown jelly, 51
Jetsam, 60
Lighthouses, 9
Lobsters, 42
    Florida spiny, 42-43
    Migration, 43
Loggerheads, 66
Mammals
    Whales, 27

Miscellaneous
    Ambergris, 26, 27
    Bioluminescence, 16
    Bottles, 11
    Cameos, 36
    Cuban rafts, 10
    El Niño, 70
    LEGO® toys, 18
    Nurdles, 18
    Rafts, 10
    Rolls Royce, 33
    Sand, 12
    Sneakers, 18, 19
    "Stoned" crabs, 12
Mollusks
    Cameos, 36
    Chitons, 21
    Elephant tusks, 14
    Flamingo tongues, 36
    Helmets, 36
    Janthina, 54
    Limpets, 21
    Nirites, 21
    Octopodes, 38-39
    Periwinkles, 21
    Slugs, 65
    Snails, 54
    Spirula, 67
    Squid, 38, 57, 67
        (see also Squid)
Moray eel, 41
Newsletters
    Drifting Seed, 27
    Beachcombers Alert!, 18
Nurdles, 18
Octocorals
    Sea pansies, 16
    Sea pens, 16
    Sea whips, 16
    Slimy sea feather, 16
Octopodes, 38-39
Parchment tube worm, 15-16
Picard, Jacques, 50
Plankton, 20, 51
Roaches, sea, 21
Sailfish Alley, 56
Salps, 52
Sand, 12
Sand dollars, 40
Sand hoppers, 21
Sargasso Sea, 58-60, 70
Sargassum
    Crabs, 65
    Fish, 63
    Shrimp, 65
    Slugs, 65
    Weed, 59

Sea beans, 55, 61-62
    Box fruit, 62
    Calabash, 62
    Coconut, 62
    Hamburger bean, 61
    Sea heart, 62
Sea cucumbers, 41
Sea frost, 14
Sea pansies, 16
Sea pens, 16
Sea roaches, 21
Sea spiders, 65
Sea stars, 40
Sea turtles, 66
Sea weeds (also see Algae)
    Sargassum, 59
Sea whips, 16, 36
Sharks
    Tiger, 66
Shells (see Mollusks)
Ships
    Atocha, 31-32
    Breconshire, 33
    Mercedes, 33
Shrimp
    Cleaner, 44
    Ghost, 13
    Mantis, 44
    Pistol, 45
    Sargassum, 65
Snails (see Mollusks)
Spirula, 67
Sponges, 37
    Black ball, 37
    Finger, 37
    Yellow tube, 37
Squid, 38, 57, 67
    Atlantic reef, 38
    Calliteuthis, 68
    Opalescent, 67
    Spirula, 67
Urchins, 40
Whale lice, 26
Whales
    Right, 26
    Sperm, 27
Worms
    Arrow, 51
    Atlantic palolo, 46
    Bearded fire, 46
    Christmas tree, 46
    Feather dusters, 46
    Honeycomb, 14
    Parchment tube, 15
    Pelagic clam, 68
    Plumed, 14
    Sea frost, 14

Talk about Cathie Katz' nature books ...

The Miami Herald: "If you'd like a guide with some personality, this is it. It's like taking a walk down the beach with a knowledgeable friend who has an unmatched enthusiasm for collecting." (Cy Zaneski)

Florida TODAY: "Realistic pen-and-ink drawings combine with well-researched, factual and entertaining text. Katz is a letter-day pilgrim at Tinker Creek, a self-taught naturalist and a genuinely gifted artist. Joy, exuberance and freshness come through in every page. She captures the significance of small things in a distinctive and individual way." (Frank Thomas)

The Ft. Lauderdale Sun-Sentinel: "[Katz'] delight in such discoveries is contagious. She wants to open eyes to an endlessly fascinating world that most of us are blind to. From myth to science to menu, the pages abound with surprises." (Margo Harakus)

The Johns Hopkins University APL NEWS: "Her books are free of the stiff scientific approach typical of most research books ... illustrated with detailed, delightful drawings." (Helen Worth)

The Melbourne Times: "[Katz] has combined history and geography along with her stories of the things the ocean waters wash in. She writes of the sea and shore birds, the turtles, fish and the vegetation of the dunes. Her illustrations are intricately and accurately drawn." (Weona Cleveland)

The Deerfield Beach Observer: "Katz unearthed a wealth of unpublished information on odd marine objects ... her skills as a technical writer have served her well as she transforms it into short, concise reader-friendly writing." (Joan Durbin)

The Castanea Botanic Journal: "Cathie's motto is *The more you look, the more you see*. Her scientifically accurate drawings and her crisp text reflect that she really sees and records very clearly..." (Dr. Charles R. Gunn, research botanist and author)

Tom MacCubbin, State Agricultural Extension Agent and author of *Florida Homegrown*, on Florida Gardening Saturday morning radio: "Every Florida resident should have Cathie's books on their coffee table where residents and newcomers can learn what's around us -- her books are fresh, informative, easy to understand and exciting."

Florida Native Plant Society: "Katz has done it again...turned the mundane into magic. Take a journey into the soul of Florida, bugs and all. This is a book for every Floridian and every visitor to Florida."

State Park Rangers: "We all love [Cathie Katz'] books and use them as reference in many of our programs and recommend them to our visitors." (Ed Perry)

Daytona News-Journal: "Packed with hundreds of black and white illustrations, her books are busy, fact-filled, silly and profound." (Mark Lane)

Florida TODAY: "The most charmingly informative books we've ever seen!" (Milt Salamon)

To order *The Nature of Florida* Books by Cathie Katz, please fill out below and mail to:

Atlantic Press
PO Box 510366
Melbourne Beach, FL 32951

Make checks payable to Atlantic Press

Quantity

☐ *The Nature of Florida's Ocean Life* .... $8.95 x _____ = $_____
   (ISBN 1-888025-11-5)

☐ *The Nature of Florida's Neighborhoods* .... $8.95 x _____ = $_____
   (ISBN 1-888025-09-3)

☐ *The Nature of Florida's Beaches* .... $8.95 x _____ = $_____
   (ISBN 1-888025-07-7)

☐ *The Nature of Florida's Waterways* .... $8.95 x _____ = $_____
   (ISBN 1-888025-08-5)

Subtotal .... $_____

Add 54¢ per book (6% Florida tax) .... $_____

Add $2.00 postage and handling for first 2 books;
$1.00 each additional book .... $_____

Total .... $_____

(Books will be mailed within one week of order.)

Mail books to:

Name:_____

Street Address:_____Apt. No.____

City:_____State____Zip_____

For discounts (orders of 10 or more books) or educational discounts, call or FAX Atlantic Press at (407) 676-5718.

cut on dotted line or copy this page

Cathie Katz has worked with The Johns Hopkins University/Applied Physics Laboratory at Cape Canaveral Air Station as their Senior Editor since 1987. From her home office in Melbourne Beach, she created of *The Nature of Florida* books. *The Nature of Florida's Ocean Life* is the fourth in this series.

Cathie is also co-editor with Dr. C. R. Gunn of *The Drifting Seed*, an international newsletter about sea beans and things that drift in the ocean. She is the founder of the International Sea-Bean Symposium held each fall in Melbourne Beach, Florida. Cathie is currently working on her next books, *Sea Soup* and *Nature a Day at a Time* for national publication in 1999.

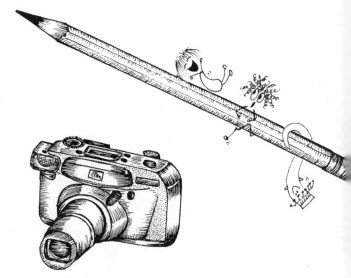

The photograph of a queen triggerfish on the front cover was taken by **Jim Angy**, long-time Florida resident and respiratory therapist. His photographs have won national awards and have been shown in Audubon's Florida Naturalist, Florida Wildlife, Popular Photography, Southern Outdoors, Orvis Outdoor catalog and on greeting cards and postcards for the Nature Conservancy. Jim is a speaker at many nature groups and wildlife organizations, sharing his expert knowledge of Florida's birds, fish, reptiles, amphibians and mammals. Jim is also the principal consultant for Atlantic Press.

"The Living Rock" on the back cover was painted by **David Williams**, a naturalist and artist originally from Lexington, North Carolina. Through his graphics business, *Wingin' It Works*, David's reputation as an accomplished wildlife artist has grown rapidly throughout the south.

David is an active member of The Drifters, an international group of beachcombers, naturalists and scientists and a presenter at the International Sea-Bean Symposium in Melbourne Beach, demonstrating scrimshaw engraving on ivory nuts and sea hearts. He also demonstrates how to create nature journals, using his own journals from the Central American rain forests as examples.

For information about obtaining a copy of "The Living Rock" or other prints by David Williams, call his pager at 336-717-1468 or write to him at 4181 Giles Road, Lexington, NC 27295.

David created a website about The Drifters at:
www.geocities.com\CapeCanaveral\Launchpad\1000

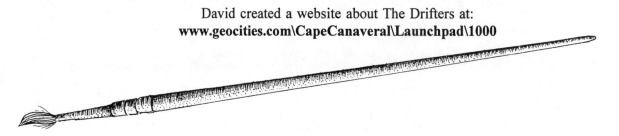

For information about the International Sea-Bean Symposium in Melbourne Beach, Florida and *The Drifting Seed* Newsletter, write (include a stamped, self-addressed envelope) to:
The Drifting Seed, PO Box 510366, Melbourne Beach, FL 32951.